Radiation from astronomical objects generally shows some degree of polarization. Although this polarized radiation is usually only a small fraction of the total radiation, it often carries a wealth of information on the physical state and geometry of the emitting object and intervening material. Measurement of this polarized radiation is central to much modern astrophysical research. This handy volume provides a clear, comprehensive and concise introduction to astronomical polarimetry at all wavelengths.

Starting from first principles and a simple physical picture of polarized radiation, the reader is introduced to all key topics, including Stokes parameters, applications of polarimetry in astronomy, polarization algebra, polarization errors and calibration methods, and a selection of instruments (from radio to X-ray). The book is rounded off with a number of useful case studies, a collection of exercises, an extensive list of further reading and an informative index.

This review of all aspects of astronomical polarization provides both an essential introduction for graduate students and a valuable reference for practising astronomers.

T0254275

ASTRONOMICAL POLARIMETRY

ASTRONOMICAL
POLAR⭥METRY

JAAP TINBERGEN

University Observatory, Leiden, Kapteyn Observatory, Roden

CAMBRIDGE
UNIVERSITY PRESS

CAMBRIDGE UNIVERSITY PRESS
Cambridge, New York, Melbourne, Madrid, Cape Town, Singapore, São Paulo

Cambridge University Press
The Edinburgh Building, Cambridge CB2 2RU, UK

Published in the United States of America by Cambridge University Press, New York

www.cambridge.org
Information on this title: www.cambridge.org/9780521475310

First published 1996
This digitally printed first paperback version 2005

A catalogue record for this publication is available from the British Library

Library of Congress Cataloguing in Publication data
Tinbergen, Jaap.
Astronomical polarimetry / Jaap Tinbergen.
p.   cm.
Includes bibliographical references and index.
ISBN 0-521-47531-7 (hc : alk. paper)
1. Polarimetry.   2. Astronomical spectroscopy.   I. Title.
QB468.T56   1996
522′.65–dc20   95-50599 CIP

ISBN-13 978-0-521-47531-0 hardback
ISBN-10 0-521-47531-7 hardback

ISBN-13 978-0-521-01858-6 paperback
ISBN-10 0-521-01858-7 paperback

To the reader who will use polarimetry as a working tool,
to my wife, who knows I have long wanted to write a book of this kind,
and to my daughter, who likes the idea that her father has a volume
of his own on the 'family shelf' in our bookcase,

this book is dedicated.

Contents

Illustrations

Tables

Preface

In het land der blinden is Eénoog koning. This saying in my mother tongue contains a sufficient number of Germanic roots for English speakers to guess that the situation depicted is only marginally better than 'the blind leading the blind'. It aptly describes the current situation in astronomical polarimetry and provides the justification for my attempt to write a primer for students and other polarimetric novices. If we can take today's students straight from polarimetric fundamentals to what is best in modern research practice, then five years from now we shall have a polarization community with both eyes wide open and firmly fixed beyond present-day horizons. That is what this book is about.

Polarimetry, performed mainly by optical or radio specialists, has already made a considerable impact on astronomy, and it deserves to be a standard observational technique, to be used whenever it is best for the job in hand. Accordingly, all astronomers should acquire polarimetric basics. My aim is to allow the reader, starting at first principles, to make use of the very latest literature. To preserve readibility, I have omitted most of the historical development. The References section at the end of the book reflects this attitude; interested readers can always trace the history backwards from modern papers.

    I have tried to resist any tendency to write a comprehensive monograph. This book is about polari*metry* and, beyond mentioning guiding principles, I have economized on astronomical applications; others are more familiar with these than I am, and the relevant literature is available in every astronomical library. The literature on the *measurement* of polarization, on polarization terminology and on mathematical methods is much more scattered, is spectrally segregated and is not very homogeneous; on these subjects, therefore, my attention was focused. Whenever possible or convenient, I have used figures from other books, reviews and research papers; this has the dual purpose of giving the reader a feeling for the (scattered) wealth already available in the literature and of forcing myself to list such work in the References section, which has gradually become one of the key parts of the book.

The book started life as a Leiden Observatory senior undergraduate lecture course, in which I divided the time roughly as follows (in units of 45 minutes duration):

chapter 1 and preview	2 units
chapter 2	6 units
chapter 3	3 units
chapter 4	4 units
chapter 5	3 units
chapter 6	4 units
chapter 7	2 units

When converting the course notes into a book, I aimed for something close to a teach-yourself primer; partly for that reason, the references are given in full format and exercises were added. In the interests of brevity, proofs are given only when they are very simple, or essential to understanding. Most readers do not need a proof of every step in a long argument, and there is no need to duplicate matters that are well covered in accessible literature. In particular, no attempt is made to include details of Fourier theory when discussing synthesis imaging, and there is no detailed discussion of measurement noise, precision or sensitivity, since these are fundamentally no different from their equivalents in 'photometry'; polarimetric addenda to such topics are mentioned, however.

The book should be read quickly, as a preparation for tackling the latest research literature, which must be the starting point for fully integrating polarimetry into astronomical practice. I can only be a general guide, my readers must find their own way; I ask them to use the book for whatever insights it may give and (remaining in style) to turn a blind eye to its defects and omissions.

Jaap Tinbergen
Roden

Acknowledgements

A number of colleagues pointed out key papers in specialized literature, clarified points in discussions, read several versions of the manuscript and corrected a few serious shortcomings. Others gave significant help in production matters. My sincere apologies to anyone whom I have unjustly but inadvertently omitted from the list, which reads: Rainer Beck, Elly Berkhuijsen, Jaap Bregman, Carlos de Breuck, Wim Brouw, Ger de Bruyn, David Clarke, Georg Comello, Robin Conway, Victor Dubrovich, Bob Fosbury, Albert Greve, Johan Hamaker, Jim Hough, Joop Hovenier, Henk van de Hulst, Vincent Icke, Philip Kaaret, Philipp Kronberg, Jan Kuijpers, Egidio Landi Degl'Innocenti, Rudolf Le Poole, Alan Marscher, Nikolaus Neininger, Rob van Ojik, Alexey Poezd, Serguei Pogrebenko, René Rutten, Rob Rutten, Ton Schoenmaker, Kees van Schooneveld, Titus Spoelstra, Richard Strom, Gerrit Verschuur, Jeremy Walsh, John Wardle, Paul Wesselius, the Cambridge Philosophical Society, Latex, Internet and CUP referees and staff.

I am grateful for all the help given, and I was pleasantly surprised by the very positive response to my plans. Remaining errors are, of course, entirely my own responsibility.

1

Introduction

Almost every issue of the leading astronomical journals includes some polarimetry, either directly or indirectly. Polarimetry as a working tool has clearly
come of age. Optical and radio techniques are most advanced, but infrared,
sub-millimetre and ultraviolet are following on rapidly, while X-ray techniques
are being developed also. There is no technical reason why astronomers should
not use polarimetry when it suits their astronomical purposes; polarimetry
often yields information that other methods of observation cannot give, and
this is the main reason why all astronomers, and today's students in particular,
should understand the basic ideas behind polarimetry.

Within the astronomical context, the degree of polarization is often low; a
few per cent is typical, though both higher and (much) lower values occur. A
polarimetric measurement is basically that of the ratio of the small *difference*
between two signals to their sum. Difference and ratio methods have been
devised to measure this small difference without systematic bias or drift errors,
but photometric noise (detector noise or photon noise of the signal itself) is
always present. To reduce this noise to the low level required for sufficiently
accurate polarimetry, considerable observing time on a large telescope is generally needed. Polarimetry should therefore not be used indiscriminately, but
only when it provides insight which other methods cannot give. Such judgment
also requires a grasp of polarimetric basics.

This book aims to create an awareness of what polarimetry can do and at
what price (in observing time, in complexity of equipment and of procedures).
To create such awareness, I must introduce polarimetric concepts and jargon.
The general plan of the book is detailed below.

The nature of polarized radiation

Astronomical signals have many of the characteristics of noise, which is roughly
the same as saying that phase information is not important. However, elec-

1

tromagnetic radiation consists of *transverse* vibrations of the electromagnetic field, and two otherwise indistinguishable vibrations can separately propagate through a medium, without change and without interfering with each other. If, in spite of the noise-like character of these signals, lasting phase and amplitude relations exist between them, the wave is said to be 'polarized' (the term is a misnomer which can be traced to Newton's light corpuscles, but it has stuck; see Clarke and Grainger (1971, appendix I)). Within astronomy, generally only part of the radiation is polarized; the remainder is unpolarized. The polarized part can be a function of wavelength, time, or direction of arrival; in general, the functional dependence on these variables will be different from that of the unpolarized part. We can express this as a functional dependence of the *degree of polarization*, in which most of the astronomical information from polarimetry resides.

These concepts need to be defined, and chapter 2 is devoted to this; it will deal with linear, circular and elliptical polarizations, both total and partial.

Astronomical situations that may lead to polarization

In a very general sense, one may state that polarization yields information about *asymmetry* or *anisotropy* inherent in the astronomical configuration. Such asymmetry may be within the source itself, or in the medium between source and observer, or both. In the case of point sources, polarization is often a good way of obtaining otherwise inaccessible information about the internal structure of the source. Common asymmetries are magnetic fields or geometric asymmetries in the distribution of scattered radiation.

The wavelength range of observable polarized radiation is from γ-rays to metre radio waves. The wavelength one uses depends on the characteristics of (the part of) the source one wishes to observe. Polarimetry in widely different wavelength ranges may be used to observe asymmetries in different parts of the same object; notable examples are the Sun, quasars and Seyfert galaxies.

Chapter 3 is devoted to an overview of polarization astronomy. Pure astronomers should take this only as an appetizer; coming from an instrumental specialist, the chapter is bound to be both incomplete and naive. I have tried throughout to refer to review articles or 'typical' modern applications, so that the way back to the origin of the research area should be clear.

Mathematical formalism and computational methods

While simple concepts and simple thoughts suffice to set the scene, more advanced polarimetry needs exact definitions and mathematical ways of handling the state of polarization at each point of the source, the instrument or the intervening medium. Polarization, once generated, can be modified, and it can be destroyed by averaging. Averaging itself is a modification, so we need a formalism to represent modification of polarization. The vector character of polarization implies that the representation of the modification must convert from vector to vector, so it will be no surprise that matrices figure prominently in the modification formalisms. There are two basic formalisms, the distinction being that one of them allows the phase of the total signal to be retained (useful for instruments and possibly for astronomical sources such as masers), while the other includes treatment of partial polarization (most of the rest of astronomy). Chapter 4 deals with these formalisms, and pictorial representations are introduced where applicable. Wavelength is not an important variable in this chapter, the basic formalism being identical from γ-ray to radio; optical-region examples will generally be used.

Techniques of measurement

Such techniques range from 'looking through a Polaroid' (or the equivalent at other wavelengths) to specialized use of time-varying wave plates and sophisticated detectors like CCDs or VLBI arrays. Though the precise form of an instrument is dictated by the wavelength range for which it is designed, some instrumental principles serve over a wide range in wavelength and the unity of instrumental methods throughout the spectrum will be stressed. Errors of measurement specific to polarimetry will be discussed in some detail. Particular categories of instrumentation will be defined, such as:

- filter polarimetry of point sources;
- imaging polarimetry;
- spectro-polarimetry;
- time-resolved (imaging, spectro-) polarimetry.

Chapter 5 deals with general principles, and chapter 6 discusses the hardware; to some extent these two chapters are interdependent and should be read together.

Case studies

Having assembled all the basic concepts, we are ready (in chapter 7) to tackle advanced literature. The best way to verify that one understands basics is to read such advanced literature at the end of a good day's work; if it still seems to make sense under those circumstances, this book can be passed on to the next reader.

History

A brief history of astronomical polarimetry is given in table 1.1. Although the basic concepts were available around 1850 through Stokes' work, technical development for astronomy started with Lyot, about 75 years later. Progress accelerated soon after 1940, and by 1980 the subject had become a mature branch of astronomical engineering.

Polarimetry as an observational technique

To set the scene for the rest of the book, I list in tables 1.2 and 1.3 typical *maximum* 'signal' levels in astronomical polarimetry and typical *minimum* errors that have been achieved; this *dynamic range* is a rough indicator of how informative one may expect polarimetry to be. The data necessarily are split according to wavelength region. The numbers quoted are from many different sources in the literature and are a rough indication only.

Table 1.1 *A few milestones in the history of astronomical polarimetry*
A more detailed chronology may be found in Gehrels (1974).

1669	Bartholinus discovers double refraction in calcite
c. 1670	(Christiaan) Huygens interprets this in terms of a spherical and an elliptical wavefront and discovers extinction by crossed polarizers
1672	Newton considers the light and the crystal to have 'attractive virtue lodged in certain sides' and refers to the poles of a magnet as an analogy; this in time leads to the term 'polarization'
1852	Stokes studies incoherent superposition of polarized light beams and introduces four parameters to describe the (partial) polarization of noise-like signals
1923	Lyot performs polarimetry of the sunlight scattered by Venus; this is the start of *polarimetry* as an effective *astronomical* technique
1942	polarization concepts and sign conventions are clearly and unambiguously defined by the Institute of Radio Engineers (IRE, nowadays IEEE); radio astronomers adopt these conventions
1946	Chandrasekhar introduces the Stokes parameters into astronomy and predicts linear polarization of electron-scattered starlight, to be detected in eclipsing binaries
1949	Hiltner and Hall set out to verify this prediction and actually find interstellar polarization
1953	Shklovskii proposes synchrotron radiation as the dominant mechanism for polarized radio emission and for the optical continuum of the Crab Nebula; Dombrowsky detects optical polarization in the latter
1957	first detection of polarized astronomical radio waves (from the Crab Nebula and, marginally, in the Galactic continuum)
1972	a team from Columbia University detects polarized X-ray emission from the Crab Nebula
1973	the International Astronomical Union (IAU; commissions 25 and 40) endorses the IEEE definitions for elliptical polarization
1974	the first source book of astronomical polarimetry is published (Gehrels 1974)
c. 1990	*common-user polarimeters* in several wavelength ranges and theoretical developments in astrophysics contribute to making polarization acceptable within mainstream astronomy
2002	'Stokes parameters; the first 150 years'. At this international conference, a joint IAU/OSA/IEEE committee 'urges authors to state the conventions they use' and refuses to enforce any particular system; chaos reigns and astronomical polarimetry flourishes

Table 1.2 *Maximum observed or expected degree of polarization*
Rough literature survey, 1994.

Radio

Galactic continuum	70%
quasars (integrated)	15%
quasars (resolved)	70%
extragalactic jets	50%
cosmic microwave background (33 GHz)	<0.01%
Crab Nebula (1 arcsec resolution)	30%
pulsars (linear)	80%
pulsars (circular, instantaneous)	70%
OH masers (as seen in VLBI)	100%
Galactic Zeeman 21 cm, 18 cm absorption	2%
solar flares, flare stars, other stars (circular)	10−100%
extragalactic (circular, integrated)	0.1%
extragalactic (circular, resolved)	0.5%

Infrared/sub-millimetre

scattering by interstellar dust grains ($1-4\,\mu$m)	75%
dust emission	2%

Optical

planets	>20%
interstellar dust acting on starlight (linear)	10%
interstellar dust acting on starlight (circular)	0.05%
Sun and A_p stars (Zeeman effect)	100%
white dwarfs (Zeeman effect)	12%
symbiotic stars (Raman scattering)	8%
reflection nebulae (including Herbig-Haro and bipolar)	60%
post-AGB stars and proto-PN (global polarization)	30%
synchrotron (Crab Nebula, blazars)	50%
synchrotron (extragalactic jets)	20%
Crab pulsar	20%

Ultraviolet

interstellar dust acting on starlight (linear)	4%
scattered light within NGC 1068	60%

X-ray (mainly 'expected')

solar flares	5%
Crab Nebula (2.6 keV)	15%
accreting X-ray pulsars	80%
rotation-powered X-ray pulsars	10%
black hole (Lense-Thirring effect Cyg X1)	2%
active galactic nuclei	20%
Seyfert accretion disc reprocessing	5%

γ-ray ('expected')

pulsars	100%

Table 1.3 *Accuracy in measurement of degree of polarization*
Rough literature survey, 1994.

	Minimum error (best ever!)
Radio	
emission (single dish; point source)	≈0.5%
emission (single dish; background)	1%
emission (interferometers, WSRT)	≈0.03%
emission (VLA, VLBA; on-axis)	≈0.2%
emission (VLBI; dissimilar telescopes)	≈2%
absorption line (Zeeman; single dish)	0.1%
absorption line (Zeeman; synthesis array)	≈0.03%
emission line (Zeeman; single dish)	0.15%
maser line (Zeeman OH VLBI)	≈1%
time-resolved (pulsar)	3%
Infrared/sub-millimetre	
1.3 mm	0.01%
270 μm	0.2%
77 μm	0.1%
20 μm	0.5% (1979)
11 μm	0.2%
1−4 μm	0.5%
Optical	
photomultiplier (stars and planets, linear)	0.001%
photomultiplier (stars and planets, circular)	0.0002%
photomultiplier (solar)	0.0001%
time-resolved (pulsar, photomultiplier)	0.2%
CCD (DC)	<0.1%
CCD (modulation; 'expected')	<0.01%
photographic	≈0.5%
visual (planets, Moon)	0.1%
Ultraviolet	
Hubble Space Telescope FOS (140–330 nm; pre-COSTAR)	≈0.2%
WUPPE on Spacelab (140–320 nm)	0.05%
X-ray ('expected' for 10^5 s)	
graphite Bragg (2.6 keV)	0.5%
lithium Thomson (5–10 keV)	0.1%
γ-ray ('expected')	
future instrumentation	5−10%

2

A description of polarized radiation

In this chapter, the main concepts of polarized radiation will be introduced and discussed. These concepts apply at all wavelengths. Electromagnetic radiation will be treated as a continuous travelling-wave phenomenon. Quantum considerations can be postponed until the moment the radiation strikes a detector and is converted into an electrical signal. Ideal detectors are not sensitive to polarization, and, to the extent that a real-life detector can be seen as an ideal one preceded by polarization optics, quantum and polarization considerations can live side by side without the one influencing the arguments concerning the other. Of the electromagnetic wave, only the electric vector will be considered; the corresponding magnetic vector follows from Maxwell's equations.

Astronomical signals are *noise-like*. These noise-like variations of electric field strength (of the electromagnetic wave) may be passed through a narrow-band filter, so that a 'quasi-monochromatic' wave remains. Such a wave contains a very narrow band of frequencies and may be seen as a sinusoidal *carrier wave* at *signal frequency*, modulated both in amplitude and phase by noise-like variations. The highest frequencies (the fastest variations) in the modulating noise determine the width of the *sidebands* around the carrier wave in the frequency spectrum. Any wide-band ('polychromatic') signal may be seen as the sum of many quasi-monochromatic signals, all with different carrier frequencies and generally each with its own amplitude and phase modulation. It might seem that the phase of such a composite noise-like signal is unimportant, certainly in astronomy where no calibration signal of absolute phase exists for reference. This simple point of view would hold for a scalar wave such as a longitudinal wave or a pressure wave. An electromagnetic wave, however, is transverse and has vector characteristics. The instantaneous electric field of the wave can be resolved into two components at right angles to each other (and to the direction of propagation). If the signal is noise-like in all respects, the two electric field components vary randomly, without any lasting correlation between them in phase or amplitude. However, if, for any frequency within the band, an amplitude and/or phase relation between the components persists for a time which is long compared to the vibration period of the wave, the resultant combined signal is less random than one might expect from pure noise, 'there is some method in the madness'. For at least part of the signal, it is then true to say that, as the two wave components pass through *a fixed point in space*, the tip of the vector that represents the instantaneous electric field of the combined wave traces out an *ellipse*, a *circle* or a *straight line*, rather than a completely random pattern. While tracing out this more organized pattern at the signal frequency, the electric field vector does vary slowly in amplitude and phase in a noise-like manner, i.e. the size (but not the shape, orientation or handedness) of the pattern varies slowly and so does the position of the tip of the vector within the pattern (at times it lags or leads a little with respect to the position it would have for strictly monochromatic radiation). The fact that such a long-term organized pattern is present within the short-term chaos is referred to as the *polarization* of a noise-like electro-magnetic wave. Corresponding to the extent to which the *flow of radiant energy* of the noise-like electromagnetic wave is represented by such a long-term organized pattern, the wave is said to be fully polarized, partially polarized or unpolarized. The shape of the pattern is specified by referring to linear, circular or (the general case) elliptical polarization.

These basic concepts will be refined and quantified in the sections that follow. The line of argument starts with an abstraction far removed from astronomy: a strictly monochromatic wave, 100% polarized. It then proceeds to quasi-monochromatic (which nature may provide in the form of line radiation from a cool low-pressure stationary source; alternatively a high-spectral-resolution instrument may select it, out of what nature offers) and finally to polychromatic, partially polarized, which is the usual type of signal met with in astronomy.

Note: There has been a great deal of confusion in the literature (and in at least this author's mind) regarding three-dimensional pictorial representations of a polarized wave. One way to represent such a wave is by showing how the tip of the electric vector varies *in time*, as the wave passes through a fixed *point* in space, two of the axes in the diagram representing field strength, the third representing time; the polarization ellipse is the projection of this on to a plane $t = t_0$. The other way is to show a *snapshot* of the instantaneous electric field vector distribution in space, two of the axes again representing field strength while in this case the third is spatial; this pattern should then be thought of as moving through space, unchanged, at the velocity of light, the intersection with $z = z_0$ tracing out the polarization ellipse. In the usual perspective drawings of these alternative elliptical helices, they are of opposite sense for the two representations.

It is instructive to read Clarke's (1974a and 1974b, pp. 47–50) comments on these alternative representations; the comments are as valid today as they were in 1974. Figure 1 of Rees (1987) is a clear illustration of the relationship between a snapshot and the polarization ellipse, provided the x- and y-axes of that figure are relabelled as E_x and E_y, respectively.

2.1 Fully or 100% polarized radiation

Linear polarization

(i) A MONOCHROMATIC LINEARLY POLARIZED wave is the simplest concept. It has a transverse electric field *with constant orientation*, its strength at any one point in space varying strictly sinusoidally with time. The duration of this wave is infinite; it has constant amplitude and frequency for all time. Good approximations in real life are light from a well-stabilized laser (very nearly monochromatic), filtered by a Polaroid, and the radiation from a dipole antenna driven by a sine-wave signal generator (stable single-frequency oscillator). The laser and the signal generator are assumed to be switched on for an infinite time.

(ii) We can conceptually convert this wave into one that is still 100% LINEARLY POLARIZED, but is QUASI-MONOCHROMATIC, by allowing the amplitude and phase to vary slowly and often randomly. The faster these 'slow' variations are, the broader will be the range of frequencies contained within the wave (as described by Fourier transform theory). If we modify the wave in no other way, it is still 100% polarized: i.e. *all of its energy* is still transported by a transverse *linear vibration* with a well-defined orientation; all we have done is to distort, slowly and therefore only slightly, the carrier wave sinusoid. Light from a filament lamp, filtered through a monochromator and a Polaroid, is a good approximation, as is the radiation from a dipole driven by a sine-wave generator with slowly varying phase and amplitude modulation, or by a radio-frequency noise source through a narrow-band electronic filter. If we rotate the Polaroid or the dipole 'very slowly' (i.e. slowly even compared to the 'slow' amplitude and phase variations), we still have 100% linear polarization of quasi-monochromatic radiation, but with variable orientation (we say that the position angle of the direction of vibration, the 'polarization angle', varies); this would not be allowed with strictly monochromatic radiation since rotation of the orientation of the polarization would modify the vibration, which therefore would not continue for infinite time and thus would no longer be monochromatic.

(iii) Fully or 100% LINEARLY POLARIZED POLYCHROMATIC radiation is a superposition of quasi-monochromatic waves of many different frequencies;

there is usually no stable phase relation between the electromagnetic field at different frequencies. Examples are light from a filament lamp filtered only through a Polaroid, or the radiation from a dipole antenna driven directly by a source of radio-frequency white noise. There is now no single dominant frequency, amplitude or phase, just a unique *orientation* of the otherwise often randomly vibrating electric field vector ('random' is in this context to be taken as: 'random within the constraints of the mean flow of radiant energy and of the spectral bandwidth').

> **Note:** No clear dividing line exists between quasi-monochromatic and poly-chromatic radiation; it is a matter of convenience. One *can* regard poly-chromatic radiation as a sine wave modulated in phase and amplitude, but the wider the bandwidth (the faster the modulation), the less useful this concept becomes and the more attractive the polychromatic description is. In practice, the fractional bandwidth is the criterion; when it is small enough that one may neglect any frequency dependence of wave amplitude, phase, receiver gain, refractive index and such (e.g. many practical lasers and some radio applications), one uses a quasi-monochromatic description, but when functional dependence on frequency is important, a polychromatic description is more appropriate.

Two *independent monochromatic linearly polarized waves,* of the same frequency but with vibration directions at right angles to each other, can propagate through empty space and other homogeneous isotropic media, along the same path and at the same time. They are both solutions of Maxwell's equations and *only two* independent solutions are possible: a linearly polarized wave of any other orientation can be seen as an in-phase combination of these two basic waves, the ratio between their amplitudes determining the position angle of the direction of vibration of the resultant. One may choose any two orientations at right angles as the base of the representation; popular choices are horizontal/vertical, right ascension/declination, latitude/longitude (ecliptic or galactic) or as dictated by the problem studied: parallel and perpendicular to the scattering plane, rotation axis of a magnetic star, symmetry axis of a double radio source, etc. In the strictly monochromatic case, the amplitude ratio of such basic polarized waves is necessarily constant for all time and the result is always 100% polarization. In the quasi-monochromatic case, the amplitude ratio may be constant or it may vary 'very slowly' without the (linear) polarization becoming significantly less than 100%.

> **Note:** Linearly polarized radiation is sometimes said to be 'plane polarized'; the term is not as common as it used to be. The 'plane of polarization' is also an old term, which in fact used to refer to the direction of vibration of the *magnetic* field. Though often used nowadays to denote the direction of vibration of the *electric* field, the term is ambiguous in several ways and it is best avoided altogether;

it is much better to refer to the 'direction of vibration (of the electric vector)'.
See Clarke and Grainger (1971, appendix I), for more detail on this point of
terminology; see also the definition of 'plane of polarization' in IEEE (1969).

Circular polarization

(i) A MONOCHROMATIC CIRCULARLY POLARIZED wave can be seen as a combination of two monochromatic linearly polarized waves with vibration
   directions at right angles to each other, of equal amplitude and differing by
   $\pm 90°$ in phase. The combined electric field vector has constant magnitude,
   but its orientation moves uniformly with time, making one revolution per
   wave period, rotating 'left' or 'right' according to the sign of the phase
   difference; all the radiant energy is associated with this circular pattern.
   Being monochromatic, the wave has infinite duration.

(ii) A QUASI-MONOCHROMATIC 100% CIRCULARLY POLARIZED wave differs from
   its monochromatic equivalent only by slow and often random variations
   of the amplitude and of the 'circular velocity'. The tip of the electric field
   vector still moves around the circle, on average at the wave frequency,
   but it now drifts around its mean position on the circle, while the size
   of the circle also changes slowly ('slowly' denoting a speed in keeping
   with the bandwidth of the signal). If one wishes, one can still regard the
   signal as the superposition of two quasi-monochromatic *linearly* polarized
   waves with 90° phase difference, now with mutually synchronized drifts
   in amplitude and phase; however, the concept of slow changes in the size
   of the circle and, generally independently of this, the drifts around its
   circumference is a much cleaner one. For the usual astronomical signals,
   the slow drifts in circle size and in position on the circle are both random.

(iii) POLYCHROMATIC 100% CIRCULARLY POLARIZED radiation is a superposition of quasi-monochromatic waves of many different frequencies, but all
   circularly polarized in the same way. There is in general no stable phase
   relation between waves at different frequencies. Alternatively, viewing the
   total signal as a phase- and amplitude-modulated carrier wave, changes
   in the size of the circle and of the position on its circumference are much
   faster than in the quasi-monochromatic case (but still 'slow', in keeping
   with the increased but always finite bandwidth of the polychromatic signal). All the radiant energy is still associated with the circular motion of
   the field vector tip, the circle having the same left- or right-handedness
   at all frequencies in the band. If one wishes to view this signal as the
   superposition of two linearly polarized polychromatic waves, the (faster,
   but still 'slow') drifts in amplitude and phase of these two components
   must be synchronized, just as for the quasi-monochromatic case.

Two circularly polarized modes are possible, and they can propagate through empty space (and other homogeneous isotropic media) independently, their electric field vectors rotating in opposite directions; they are referred to as left-hand-circular (LHC) and right-hand-circular (RHC), although much confusion exists as to which of the two shows clockwise rotation of the electric field vector and about the point of view for examining the circle (facing the source or looking along the direction of propagation). See Simmons and Guttmann (1970) and Clarke (1974a,b) for discussions of the sign conventions and nomenclature; see also fig. 2.1.

As noted, LHC and RHC may be seen as combinations of two linearly polarized waves of equal amplitude with +90° or −90° phase difference. However, LHC and RHC waves may themselves be used as the base for other polarization forms in the same way as two linear vibrations: two *phase-correlated* circularly polarized waves of equal amplitude add to give linear polarization, the orientation of which depends on the (constant) phase difference between the circular constituents (position angle of the direction of linear vibration = 1/2 phase difference).

Note: Linearly polarized radiation has a vector character: one must specify the orientation as well as the 'intensity' (flow of radiant energy) of the radiation. Circularly polarized radiation, however, requires only a single scalar, which can specify both the 'intensity' and the 'sense' (or 'handedness', LHC or RHC, specified by the *sign* of the scalar quantity). We shall encounter this distinction again in chapter 3, in relation to the kind of asymmetry causing the polarization.

Elliptical polarization

(i) The most general form of polarization is *elliptical*, for which the tip of the electric field vector executes an ellipse at the signal frequency. The distinguishing parameters of the ellipse are *orientation, axial ratio* and *handedness*; linear and circular polarization are special cases of this general form. A MONOCHROMATIC ELLIPTICALLY POLARIZED wave may be visualized as the sum of two *unequal* linearly polarized components with phase difference of ±90°, or as the sum of two linearly polarized components (which may or may not be equal) with a phase difference of something other than 0° or ±90°. This exercise in geometry is left to the reader. It is also possible to see elliptical polarizations as the sum of two *unequal* oppositely circularly polarized components (or even one circular and one linear component, though such a 'non-orthogonal' form is less useful) with a constant phase difference between them; these useful mental gymnastics are also left to the reader.

(ii) QUASI-MONOCHROMATIC 100% ELLIPTICALLY POLARIZED radiation is obtained by allowing the size of the ellipse to vary slowly, with similar slow

variations of the position of the tip of the electric field vector within the
ellipse (lag or lead with respect to the monochromatic equivalent). The
elliptical pattern, however, is a constant of the wave. Alternatively, one
may introduce (correlated) slow and often random variations of amplitudes and phases of the two linearly polarized waves which were used in
the mental picture to construct the elliptically polarized wave.

(iii) POLYCHROMATIC 100% ELLIPTICALLY POLARIZED radiation is a sum of quasi-
monochromatic components, all with the same elliptical polarization. Its
noise-like character is entirely analogous to the case of circular polarization discussed above. In spite of all the amplitude and phase variations,
the ellipse is again a constant of the wave.

Elliptical polarization is the most general form possible, linear and circular
polarization being special cases; linear polarization has axis ratio 0, circular
polarization has axis ratio 1. We have mentioned that two independent linear or
circular polarization forms can be sustained in homogeneous isotropic media,
these forms having 'opposite' characters to each other: two linear forms with
directions of vibration at right angles, two circular forms of opposite handedness. By considering these limiting forms, one may suspect that 'opposite'
elliptical polarization forms must have equal axial ratio, that the major axes of
the two ellipses must be at right angles to each other and that the ellipses must
be traced out in opposite directions (opposite handedness). This statement will
be quantified in section 2.3.

The position angle of the axes of the ellipse may be changing 'very slowly'
with time, provided the radiation is quasi-monochromatic or polychromatic;
similarly, the ellipticity of the ellipse may be changing 'very slowly'. This
is entirely analogous to the case of a slowly rotating Polaroid discussed for
quasi-monochromatic linear polarization on p. 10.

2.2 The Stokes parameters

We now introduce the Stokes parameters, four quantities which all denote
'radiant energy per unit time, unit frequency interval, unit (detector or collector)
area (and for extended sources: per unit solid angle)'; see 'Note on units' on
p. 16. This representation of polarized light was invented by Sir George Gabriel
Stokes (1852); it was revived and introduced into astronomy by Chandrasekhar
(1946). The absolute phase of the wave does not enter into the definition;
addition of the Stokes parameters of beams of radiation represents incoherent
superposition of these beams. The Stokes parameters are often gathered into
a 4-vector $\mathbf{S}$ with components labelled S_0, S_1, S_2, S_3; or I, Q, U, V (which will be

Polarization ellipse   **Stokes parameters**

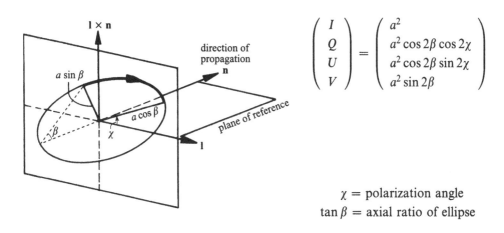

$$\begin{pmatrix} I \\ Q \\ U \\ V \end{pmatrix} = \begin{pmatrix} a^2 \\ a^2\cos 2\beta\cos 2\chi \\ a^2\cos 2\beta\sin 2\chi \\ a^2\sin 2\beta \end{pmatrix}$$

χ = polarization angle

$\tan\beta$ = axial ratio of ellipse

Fig. 2.1 The Stokes parameters for 100% polarized radiation, adapted from Van de Hulst (1980, p. 495). In the convention recommended in 1942 by the Institute of Radio Engineers (IRE, now IEEE) and endorsed in 1973 by the relevant commissions of the International Astronomical Union (IAU), the polarization shown is right-handed elliptical and V is positive (Kraus 1966, table 4.1 and its footnote; Simmons and Guttmann 1970, appendix III; IAU 1973; Conway 1974, footnote p. 353). The unit vectors $\mathbf{l}$ and $\mathbf{l}\times\mathbf{n}$ relate the present figure to figs. 2.2 and 2.3.

used in this book); or I, M, C, S. In terms of the properties of the polarization ellipse, they are defined by the equations shown in fig. 2.1. Other, equivalent, definitions exist; they can be found in section 4.3. Some of these are more elegant, or are particularly useful for defining exactly what a polarimeter should measure. However, for handling noise-like signals in which phase is irrelevant but a 'polarization ellipse' is a suitable description for the method in the madness, the form of fig. 2.1 is appropriate: electric field amplitudes are squared, phase does not appear and, finally, the parameters of the polarization ellipse appear in the definition more or less in the functional form one might expect. The full impact of the Stokes parameter representation will gradually be appreciated as we use it for practical purposes. Let us for the moment accept Sir George's gift horse in good grace; we can decide on its bite when we have got to know it better.

Note that $I \geq 0$, but Q, U, V can be positive or negative. Note also that the double angles 2β and 2χ enter into the definitions; the basic reason for this is the squaring involved in going from amplitude to radiant energy ('intensity').

Let us examine the form of the Stokes parameters for 100% linear, 100% circular and 100% elliptical polarization (fig. 2.1). For LINEAR polarization, $\sin \beta = 0$ or $\cos \beta = 0$. In both cases $\sin 2\beta = 0$, so $V = 0$, while $Q = a^2 \cos 2\chi$, $U = a^2 \sin 2\chi$. The quantity a is related to the amplitude of the electric field vibration (so a^2 is concerned with 'intensity', or flow of radiant energy), the angle χ is the orientation of the ellipse (in this case of the straight line) with respect to the chosen reference direction; χ is called the 'azimuth' of the polarization, or the 'polarization angle'. The quantities Q and U are Cartesian components of the vector $(a^2, 2\chi)$; note the doubling of the polarization angle in this true vector representation. A Q, U-diagram (or $Q/I, U/I$) is one of the common representations in astronomical polarimetry. Note that for the polarization ellipse the origin of χ is chosen arbitrarily (e.g. towards zenith, equatorial North Pole or Galactic North Pole, along the symmetry axis of a double radio source), but in the Q, U-diagram the origin of 2χ is by definition the Q-axis (which is conventionally drawn horizontally).

For CIRCULAR polarization, $\sin \beta = \pm \cos \beta$ or $\sin 2\beta = \pm 1$, $Q = U = 0$ and $|V| = a^2 = I$. For the sign conventions for V, see Kraus (1966, table 4.1 and its footnote), Simmons and Guttmann (1970, appendix III) and Clarke (1974a,b). Given the apparently contrary conventions of radio astronomers (Conway 1974, footnote p. 353) and 'traditional' optical astronomers (Rees 1987, p. 216), I hesitate to recommend either. Although a convention is officially recommended by the International Astronomical Union (IAU 1973), the main point to note is that sign conventions in astronomical polarimetry are many and varied, and that one should be particularly careful in specifying one's own choice in all future research papers.

General ELLIPTICAL polarization is represented by non-zero values of Q, U and V. Note that for 100% polarization (which is all we have mentioned so far) $Q^2 + U^2 + V^2 = I^2$. The axial ratio of the ellipse is $\tan \beta$.

Note on units: I shall side-step the hornets' nest of radiometric quantities and units (magnitudes, candela per foot^2-angstrom, jansky, counts per second, etc.), merely noting that in astronomy we deal with what in radiometry is called *radiant intensity* for point sources and *radiance* for extended sources (the difference being whether one integrates over the source solid angle or not). Questions of units are for photometrists; polarimetrists as such fortunately deal with relations between the four Stokes parameters, *all of which* (in any one application) represent similar physical quantities and are measured in the same units. Irrespective of the radiometric quantity being discussed or the radiometric units being used, the term *(total) intensity* is often used in polarimetry for Stokes I, and *polarized intensity* for Stokes Q, U and V or some combination of these. Readers are warned of this usage and are urged to be more specific in their own publications. Refer to Snell (1978) for an excellent overview of radiometric quantities and units; an overview of astronomical usage is given by Léna (1988, section 3.1). Berkhuijsen (1975) clarifies early confusion in radio-polarimetric terminology.

Why Stokes parameters? It may seem that the Stokes parameters, rather than following naturally from our physical concept of polarized radiation, are pulled out of a hat; the author sympathizes with this view, but notes that this holds to some extent for each of the alternative forms presented in section 4.3. Stokes himself (1852) used a long derivation starting from time-varying electric fields (transverse 'ethereal displacements') to arrive at the definition of A, B, C, D (equivalent to, respectively, I, V, Q, U of today) in the form given in fig 2.1. He then says:

Suppose that there are any number of independent polarized streams mixing together; let the mixture be resolved in any manner into two oppositely polarized streams, and let us examine the intensity of each.... It follows that if there are two groups of independent polarized streams which are such as to give the same values to each of the four quantities A, B, C, D, if the groups be resolved in any manner whatsoever, which is the same for both, into two oppositely polarized streams, the intensities of the components of the one group will be respectively equal to the intensities of the components of the other group. Conversely, if two groups of oppositely (sic!) polarized streams are such that when they are resolved in any manner, the same for both, into two oppositely polarized streams, the intensities of the components of the one group are respectively equal to the intensities of the components of the other group, the quantities A, B, C, D must be the same for the two groups.... Two such groups will be said to be equivalent.

Stokes' own formulation firmly connects (via A, B, C, D) the geometrical properties of the ellipse to experimental measurements:

It follows...that no partial analysis of light, such, for example, as would be produced by reflection from the surface of glass or metal, or by transmission through a doubly absorbing medium, can from equivalent groups produce groups which are not equivalent to each other; and we have seen already that this cannot be done by means of the alteration of phase accompanying double refraction. It follows, therefore, that equivalent groups are optically undistinguishable.

This is what makes the Stokes parameters such a supremely useful tool in describing polarized radiation. A modern derivation in the same spirit as that of Stokes may be found in Collett (1993, chapter 4).

2.3 Orthogonal modes and birefringence

For every polarization form (or 'mode'), we can define one with the same I but opposite Q, U, V; adding $\pm 90°$ to β is the operation required (fig. 2.2). For such an opposite form, the axial ratio of the ellipse is the same as that of the original, but the axis is at right angles and the ellipse is traced out in the opposite direction. Such pairs of opposites are said to be 'orthogonal' to each other; they are independent solutions of Maxwell's equations, they

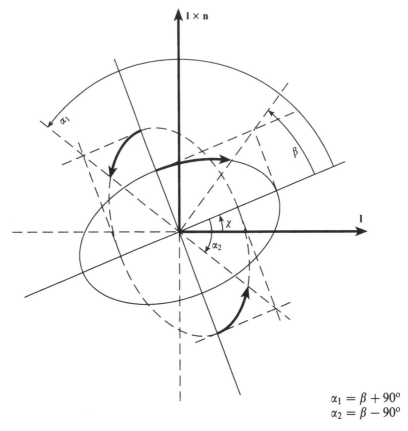

$$\alpha_1 = \beta + 90°$$
$$\alpha_2 = \beta - 90°$$

Fig. 2.2 Orthogonal polarization forms (general case). The general right-handed polarization ellipse of fig. 2.1 is shown, together with the orthogonal form obtained by inverting the last three Stokes parameters. Such inversion is accomplished by replacing β by $\alpha = \beta \pm 90°$. The axial ratio of the orthogonal form is $\tan \alpha = -\cot \beta$, i.e. the ellipse has the same shape, but its orientation has changed by 90°. Since the circular component V has been inverted, the orthogonal form is left-handed. The unit vectors l and $l \times n$ relate the present figure to figs. 2.1 and 2.3.

can propagate independently through empty space and other homogeneous isotropic media.

In a homogeneous isotropic medium, all polarization modes have the same propagation velocity. In astrophysical magnetized plasmas, however (solar active regions, pulsars, the interstellar medium, the Earth's ionosphere), different polarization modes have different propagation velocities. For a given plasma and a given direction of propagation with respect to the magnetic field, there are always two orthogonal modes that can propagate through the medium without changing their polarization form (they are 'eigenmodes'; the terms 'normal modes' and 'characteristic waves' are also used). Although the polarization of these two modes remains unchanged, they do travel at different velocities, i.e. the medium has two refractive indices, one for each eigenmode; such a medium is said to be *birefringent*. Optical crystals are also birefringent media; the astronomical significance of this lies in their use for constructing accurate polarimeters.

If a medium is LINEARLY BIREFRINGENT, its eigenmodes have linear polarization. Radiation that has exactly the polarization of one of those two modes will not be changed, but for any other polarization angle, or for circular polarization, the polarization form will change as the radiation passes through the medium. One may think of the incident radiation as being resolved into the two eigenmodes, which propagate independently, each with its own velocity. On emerging from the medium, the components recombine with a phase difference induced by the medium; the exit polarization will thus be different from the polarization of the beam that went in. In optical polarimetry, one makes use of slices of linearly birefringent crystal materials, so-called wave plates (e.g. 'quarterwave plate' for a phase difference of 90°, used for converting linear polarization to circular or vice versa).

CIRCULAR BIREFRINGENCE causes relative phase shifts between two circularly polarized eigenmodes. A linearly polarized signal impinging on such a medium should be thought of as being resolved into the two circular eigenmodes, each of which passes through the medium at its own velocity (i.e. with its own refractive index). On emerging from the medium, the two modes recombine to give linear polarization again, but the direction of vibration has been rotated by an angle of half the differential phase between the modes. In an ionized plasma with a magnetic field component along the line of sight, such rotation of the linear polarization is called 'Faraday rotation'. Circular birefringence also occurs structurally in certain optical crystals (e.g. quartz), due to a helical atomic arrangement. Such crystals can be used to construct circular retarders (or 'rotators', viz. of the direction of vibration of linear polarization). Many asymmetric organic molecules in solution cause the medium to be circularly birefringent or 'optically active'; this is used in laboratory techniques such as saccharimetry, but so far it has not been of professional significance to astronomers.

In general, birefringence will be ELLIPTICAL, i.e. the eigenmodes are elliptical. This occurs in astrophysics (e.g. Jones and O'Dell 1977), but is not used much in technical work (at least not intentionally; oblique rays through crystal components designed for normal incidence inevitably have eigenmodes with some ellipticity).

Note: Rays with orthogonal states of polarization are sometimes denoted by 'o' for ordinary and 'e' (or 'x' in radio work) for extraordinary. This usage arose from polarization effects in crystals, such as calcite, in which the extraordinary ray is not refracted in the same way as in 'ordinary' isotropic materials, and it has been adopted for the description of radio wave propagation (IEEE 1969). Nowadays, the terms o- and e-ray may loosely denote any pair of rays of orthogonal elliptical polarization, and may be encountered in descriptions of instruments (figs. 6.3 and 6.5) and of magneto-ionic plasmas, such as the Earth's ionosphere or the Sun's corona; for linear polarization, 's' and 'p', or $\perp$ and $\parallel$ are also found (fig. 5.7 and section 6.1.2).

2.4 Unpolarized radiation

What happens when we combine the electric fields of two equal (quasi-monochromatic or polychromatic) 100% polarized waves of orthogonal polarization forms, *there being no long-term persistence in the phase relation between them* ('incoherent sum', or 'intensity superposition')? For the sake of a clear and simple mental picture, we take as an example two linear polarizations at right angles. For a time short compared to the 'slow' variations of amplitude and phase discussed in previous sections, some definite polarization form will result (linear, circular or elliptical, depending on the momentary phase difference). However, a sufficiently long time average is part of the definition of the Stokes parameters (Stokes himself is quite clear about this, distinguishing between 'temporary intensities' and 'actual intensities'). After some time has elapsed, the 'slow' variations will have caused the polarization form of the incoherent sum to change to something else. During a 'sufficiently' long time interval, all possible forms of polarization will occur, all values of β and χ will occur, and the time-averaged Stokes parameters will be $<a^2>, 0, 0, 0$, where $< >$ denotes a time average. We call such radiation *unpolarized*, since no single polarization form dominates or is conspicuously absent. We note that the Stokes parameters of the two linear polarizations that went into the incoherent sum were $I/2, Q, U, 0$ and $I/2, -Q, -U, 0$, and that for such intensity superposition the Stokes parameters of the sum are equal to the sums of the Stokes parameters of the components. (A mathematically inclined reader may prefer the proof of additivity of the Stokes parameters by, e.g., Collett (1993, section 4.6).)

The same argument may be used for *incoherent* addition of equal LHC and RHC polarized components. Again, the sum has $Q = U = V = 0$. Unpolarized radiation can be seen as the incoherent sum of two beams, of any two 'orthogonal' polarizations $I/2, \pm Q, \pm U, \pm V$. It does not matter what polarization forms one chooses, as long as the Stokes vector sum is $I, 0, 0, 0$, in other words as long as the last three Stokes parameters are equal but opposite in the two components.

Parallelling the above definition of unpolarized radiation in terms of Stokes parameters, one may define it in operational terms, i.e. in terms of hypothetical measurements. One may set up 'polarimeters' (instruments that are sensitive to and measure some particular form of polarization) for any polarization form one likes to choose. If the measurement yields a null result *no matter what polarization form one tries to detect*, the radiation under scrutiny is said to be unpolarized. The first form of the Stokes parameters in section 4.3 expresses this approach.

Note 1: Stokes, in 1852, was well aware of the equivalence of the above two points of view on unpolarized radiation (he refers to unpolarized radiation as *common light*; this term has gone out of use):

The experimental definition of common light is, light which is incapable of exhibiting rings of any kind when examined by a crystal of Iceland spar and an analyzer, or by some equivalent combination [such as a modern 'polarimeter' consisting of Babinet compensator, linear polarizer and a detector—JT]. *Consequently, a group of independent polarized streams will together be equivalent to common light when, on being resolved in any manner into two oppositely polarized pencils, the intensities of the two are the same, and of course equal to half that of the original group. Accordingly, in order that the group should be equivalent to common light, it is necessary and sufficient that the constants B, C, D should vanish.*

Note 2: Unpolarized radiation is sometimes called *natural* radiation, but like 'common light' this term is ambiguous and is best avoided.

Note 3: Instead of averaging over time, one could use ensemble-averaging to define the polarization of noise-like signals. In fact, quasi-monochromatic and polychromatic radiation will usually be the result of ensemble-averaging of microscopic events, such as emission of quanta of line or continuum radiation, scattering of light by an ensemble of molecules, or pulses of synchrotron emission from an ensemble of relativistic electrons. In practice, therefore, averaging over time is more or less equivalent to ensemble-averaging, though in special cases (e.g. very fast time variations of polarization of light from a star) one may have to be more careful about the definitions one uses.

Note 4: Stokes uses the term 'independent' to mean *from separate sources*, i.e. without any correlation in the 'slow' variations of phase and amplitude. In my own text of this chapter, 'independent' is used only in the sense of *independent solutions of Maxwell's equations*, which is something different.

2.5 Partial polarization

Now that we have defined unpolarized radiation, the concept of partial polarization is simple: partially polarized radiation is the incoherent sum of an unpolarized and a fully polarized component. The Stokes parameters of partially polarized radiation are the sums of the Stokes parameters of the components; the I values (always positive) add, while the Q, U and V values are those of the fully polarized component. Therefore, for partially polarized radiation, $Q^2 + U^2 + V^2 < I^2$. We call $\sqrt{(Q^2 + U^2 + V^2)}/I$ the *degree of polarization*, generally denoted by p. One also encounters the degree of linear polarization $p_{\text{lin}} = \sqrt{(Q^2 + U^2)}/I$, degree of circular polarization $p_{\text{circ}} = V/I$, or other, often historically determined, forms (e.g. $p_s = -Q/I$ in optical planetary polarimetry, where $U \equiv 0$ for proper choice of coordinate frame: subscript s for 'sign'; also p_x for Q/I, p_y for U/I).

Equally valid is a representation of partially polarized radiation as the incoherent sum of two *generally unequal* fully polarized components of orthogonal polarization (in this view, unpolarized radiation is just a special case of partial polarization). I shall mention this point of view again in section 4.2.

In terms of the elliptical envelope of the electric field vector, one can visualize the field vector of partially polarized radiation as 'slowly' making the transition from one elliptical envelope to another, with some ellipses occurring more frequently than others. More radiant energy is transported by one particular elliptical type of vibration than by any of the others; note that phase does not occur in this statement.

For obvious reasons, fully or 100% polarized radiation is sometimes said to be in a *pure* state of polarization; similarly, partially polarized and unpolarized radiation, as an incoherent sum of two pure states, may be referred to as being in a *mixed* state of polarization.

2.6 The Poincaré sphere

The Poincaré sphere (fig. 2.3) is a very useful pictorial representation of polarization. It can deal with partial polarization, so most astronomical situations can be represented by it, and the operation of 'optical' components on the polarization of radiation can be represented as operations on or within the sphere. Its main use is as a graphical device, but some instrumental components have actually been designed by spherical trigonometry on the surface of the Poincaré sphere.

- On the *equator*, $2\beta = 0$, so polarization is linear, with 2χ the parameter representing the orientation.

- The *poles* represent circular polarization, and the rest of the *surface* of the sphere represents various forms of elliptical polarization.

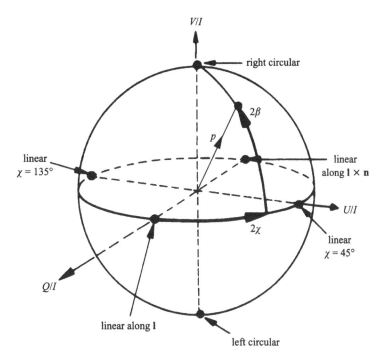

Fig. 2.3 The Poincaré sphere, adapted from Van de Hulst (1980, p. 495). The unit vectors **l** and **l** × **n** relate the present figure to figs. 2.1 and 2.2.

• Fully polarized radiation of one form or another is represented on the *surface*, the *centre* corresponds to unpolarized radiation, and all other points *within the sphere* represent partial polarization, the *length of the radius vector* representing the degree of polarization p.

The equatorial plane with axes $Q/I, U/I$ (often denoted as p_Q, p_U or p_x, p_y or q, u or even just Q, U) is the part of the Poincaré sphere most often met within astronomy; this is because partial linear polarization is by far the most common form in the universe. A diagram in this equatorial plane is a true vector diagram, as opposed to a map of polarization *lines* (which have no sense, just orientation; the polarization lines are often called vectors, however, and a map of them may erroneously be called a vector map, so beware of such confusion). The true vector diagram is often called a 'Stokes plot', which is an unambiguous term as long as only linear polarization is considered. Occasionally, the equatorial plane is presented as the complex plane, with the U-axis being imaginary. In that case, $m = Q/I + iU/I = pe^{2i\chi}$ with m referred to as the 'complex degree of polarization' (in a few papers called 'Stokes vector'; this is exceedingly confusing, since 'Stokes vector' usually denotes the 4-vector I, Q, U, V).

Table 2.1 *Representative Stokes vectors, after Shurcliff (1962)*

Pattern	L/C/E/U l/r	χ (deg)	$\tan\beta$	A_y/A_x	$\phi_y - \phi_x$ (deg)	I, Q, U, V
—	L	0	0	0	–	1, 1, 0, 0
|	L	90	0	∞	–	1, −1, 0, 0
/	L	45	0	1	0	1, 0, 1, 0
\	L	−45	0	1	±180	1, 0, −1, 0
/	L	any	0	> 0	0 or ±180	1, cos 2χ, sin 2χ, 0
◯	C,r	–	1	1	90	1, 0, 0, 1
◯	C,l	–	1	1	−90	1, 0, 0, −1
⬭	E,r	0	0.5	0.5	90	1, 0.6, 0, 0.8
⬭	E,l	0	−2	2	−90	1, −0.6, 0, −0.8
⬭	E,r	90	2	2	90	1, −0.6, 0, 0.8
⬭	E,r	45	$\tan\beta$	1	2β	1, 0, cos 2β, sin 2β
⬭	E,r	22.5	0.318	0.518	45	1, $\sqrt{1/3}$, $\sqrt{1/3}$, $\sqrt{1/3}$
-	U	–	–	–	–	1, 0, 0, 0

The polarization form is represented by χ and β, or by the alternative of x and y amplitude ratio A_y/A_x and phase difference $\phi_y - \phi_x$ (cf section 4.2). L/C/E/U denotes linear/circular/elliptical/unpolarized and l/r denotes left-handed/right-handed; right-handed corresponds to $V > 0$.

2.7 Further reading and some more discussion

Examples of various polarization forms and the corresponding Stokes vectors are shown in table 2.1. More examples may be found in the books listed below. The reader is urged to verify the examples in the table and to construct more. The Stokes parameters do have some unusual properties, and an effort must be made to understand these, for we shall use the Stokes parameters in chapter 4. For further reading, I recommend the following books; they all concern optical polarimetry, but this is of no importance for the discussion of the Stokes parameters.

The most compact yet clear description of the Stokes parameters and the Poincaré sphere that I know of is in Van de Hulst (1980, pp. 494–6). Hecht and

Zajac (1974) and Kliger *et al.* (1990) provide excellent general introductions to polarization concepts; Kliger *et al.* (1990, pp. 24–6) are clearest on pictorial representations of circularly polarized radiation. Another well-written classic is Shurcliff (1962). Collett (1993) is a wide-ranging monograph, very informative on the history of the Stokes parameters and how they relate to electromagnetic theory and Maxwell's equations. Despite it being the most expensive by far, I consider Collett (1993) the best single buy for those who wish to go well beyond the brief introduction I have given.

Note: The author finds that the following thought experiment helps him to understand the significance of the Stokes parameters. The reader is invited to try it, too: if you get lost, blame the author and press on. J.W. Hovenier clarified an essential didactic point.

Let us consider the normalized (i.e. $I = 1$) Stokes vector $1, q, u, v$, with degree of polarization $p = \sqrt{q^2 + u^2 + v^2}$. We can look on this as the sum of a 100% elliptically polarized part and a remainder, which is unpolarized:

$$\begin{pmatrix} 1 \\ q \\ u \\ v \end{pmatrix} = \begin{pmatrix} p \\ q \\ u \\ v \end{pmatrix} + \begin{pmatrix} 1-p \\ 0 \\ 0 \\ 0 \end{pmatrix}$$

For the polarized part, the parameters of the ellipse are: $\tan 2\chi = u/q$ and $\tan 2\beta = v/\sqrt{q^2 + u^2}$.

We may also split the vector into a 100% circularly polarized part, a 100% linearly polarized part and an unpolarized remainder:

$$\begin{pmatrix} 1 \\ q \\ u \\ v \end{pmatrix} = \begin{pmatrix} \sqrt{v^2} \\ 0 \\ 0 \\ v \end{pmatrix} + \begin{pmatrix} \sqrt{q^2 + u^2} \\ q \\ u \\ 0 \end{pmatrix} + \begin{pmatrix} 1 - \sqrt{v^2} - \sqrt{q^2 + u^2} \\ 0 \\ 0 \\ 0 \end{pmatrix}$$

In both these Stokes vector equations, the components of the radiation are supposed *incoherent* with each other, or Stokes vector addition would not apply. The first representation is always possible, since $p \le 1$ and the unpolarized part at worst reduces to zero. In the second representation, however, the intensity of the unpolarized part can become negative (a physical impossibility) if

$$\sqrt{v^2} + \sqrt{q^2 + u^2} > 1$$

or

$$|p_{\text{circ}}| + p_{\text{lin}} > 1$$

In a $|p_{\text{circ}}|, p_{\text{lin}}$ diagram, the limit $|p_{\text{circ}}| + p_{\text{lin}} = 1$ is a straight line from (0,1) to (1,0). The fundamental limit on $|p_{\text{circ}}|$ and p_{lin} is that of 100% polarization:

$$p_{\text{circ}}^2 + p_{\text{lin}}^2 = 1$$

In the $|p_{circ}|, p_{lin}$ diagram, this is represented by a quadrant of a circle centred on the origin and passing through (0,1) and (1,0). In the area between the line and the circle quadrant, only the first representation is allowed. In that part of the diagram it is *not* possible to represent elliptically polarized light as the *incoherent* sum of a circular and a linear component; within the triangle enclosed by the straight line and the axes, it is.

How should we interpret this physically? Remember that, if we conceptually split elliptical polarization into a circular and a linear component, these two components must be *coherent* with each other, i.e. they must perform the same random 'slow' variations of amplitude and phase; for a certain size of the circular component, the 'remainder' will contain a phase-correlated component of the right size and of the correct linear polarization to make up the original elliptical. The linearly polarized component of the second representation, however, must have a phase which is uncorrelated with that of the circularly polarized component. Hence, we conceptually take it from what was the unpolarized part in the first representation (a straight swap of the correlated for an uncorrelated component of the same polarization). If the degree of polarization is low enough, this will be possible. In the second representation, the 'unpolarized' part will therefore include a component correlated in phase with the circularly polarized part, but this phase is slowly variable and uncorrelated with the phase of the rest of the unpolarized component, so that the 'linearly polarized part of the elliptical polarization' can indeed be conceived as part of that unpolarized part of the radiation. However, when the phase-correlated component is too large to be accommodated as part of the unpolarized last term, the second representation is invalid. At low partial polarizations, there are an infinite number of possible combinations involving linear, circular and/or elliptical polarizations; usually there will be no physical reason for preferring any of these to the simplest interpretation (the first Stokes vector equation above). The Stokes representations do *not* tell us that, within the 'unpolarized' part, there may be a component which is coherent with, say, the circular component. That is as it should be: *we* have chosen to represent the total radiation in terms of (assumed) mutually incoherent components, which may not correspond at all to what nature provides; if a wrong assumption yields an incomplete answer, we have only ourselves to blame.

The pulsar illustrated in fig. 3.12 has been found to flip its polarization angle by 90° when its sense of circular polarization is reversed (quoted in Taylor and Stinebring (1986, p. 309)). If the degrees of linear and circular polarization are large enough (this cannot be decided from the figure), there is only one way of interpreting this: jumps from one elliptical mode to the orthogonal one. For a lower degree of polarization, this will still be the simplest interpretation, but alternatives in terms of incoherent linear and circular polarization, for some reason making mode jumps at the same moment, remain possible (as would other similar combinations).

3

Polarization in astronomy

In this chapter I shall discuss the scientific reasons for measuring the polarization of astronomical signals. The central question is: 'What does nature express as polarization rather than as some other property of the signal?'. This, of course, is the *scientific* point of departure for all astronomical polarimetry, but the basic concepts of polarization and (un)polarized radiation needed clarification before scientific necessity could be discussed properly. This chapter will be only a brief overview of the relevant astronomy; a number of recent reviews are available to help the reader become familiar with the astronomical applications. The subject of this book is polari*metry*, the desirability of measuring the polarization will be taken for granted.

The light of most stars is itself unpolarized. In fact, whenever one needs an optical 'zero-polarization' reference source, one is generally pushed to use stars rather than lamps. The reason for the low polarization is the great distance (point source) and the spherical symmetry of most stars: any linear polarization there might be is averaged out over the star's visible disc. In the radio domain, antenna properties are highly polarization-dependent, and without specialized techniques large spurious *apparent* polarization is generated within the instrument. Thus, circumstances conspired to make astronomical polarimetry a late arrival. Even in the spectral regions of greatest instrumental sophistication, polarimetry remained a specialist technique; solar physics has been the notable exception. As a corollary of this lack of attention to polarimetry, awareness of polarization-induced photometric errors within telescopes and instruments has been minimal.

During the last few decades, however, progress by polarization specialists has been considerable, and astronomers now realize that, even as 'common users', they neglect polarization at their peril: wherever there is appreciable *asymmetry* in an astronomical situation, there is likely to be polarization at *some* level (how this level compares with state-of-the-art accuracy of measurement is another matter). The higher the resolution (spatial, spectral and sometimes temporal), the larger in general is the polarization; the increased resolution available to astronomers is another reason for the increasing use of polarimetry in recent years.

The asymmetry notion can be pushed a little further: the character of the asymmetry determines the kind of polarization to be expected. If the asymmetry is of a scalar (+/−, magnitude) kind (e.g. the longitudinal component of the magnetic field), the polarization generated or the birefringence (section 3.3) will be circular (scalar in character). If the asymmetry is of a vector type (magnitude and direction), the polarization generated, the dichroism or the birefringence will be linear (magnitude and orientation; almost, though not quite, a vector, since it transforms into itself on rotation by 180° rather than 360°); one example of such asymmetry is the transverse magnetic field component, another is the position angle of a scatterer with respect to the primary source of the radiation.

> **Note:** The *macroscopic* argument of asymmetry is not enough to predict measurable polarization with any certainty; it only provides a hunch that polarization may occur. One also needs a *microscopic* argument in the form of a viable mechanism that will imprint this asymmetry on to the radiation we receive.

The main asymmetries giving rise to astronomical polarization are magnetic fields and an asymmetric distribution of scattered radiation. At what wavelengths these manifest themselves is partly a question of where in the spectrum the objects radiate, and this in turn can determine the sensitivity to the required data (e.g. a 10 G* field is 'very weak' for Zeeman detection on a star in the optical, whereas $10\,\mu G$ can be detected in a cold cloud by Zeeman methods at 21 cm; the zero-field width of the spectral line is the determining factor).

Specific references to reviews or the latest literature are given where possible; when this fails, try Lang (1980). For predictions of expected polarization in the fast-developing field of X-ray polarimetry, consult Kaaret *et al.* (1989) and Mészáros *et al.* (1988). For a discussion of astrophysical electric fields as a possible source of polarization effects, see Favati *et al.* (1987).

3.1 Magnetic fields and generation of polarized radiation

Polarimetry is the most direct method of detecting magnetic fields, which in astronomy range from $10\,\mu G$ in interstellar space to perhaps $10^{13}\,G$ in pulsars. The techniques used and the data produced vary according to object and wavelength. A list of applications such as that which follows will never be complete, but it can serve as a general guide.

---

* The astronomical literature still uses the gauss as the unit of magnetic flux density (magnetic induction, B-field): $1\,G = 10^{-4}\,T$ (see also Crangle and Gibbs 1994).

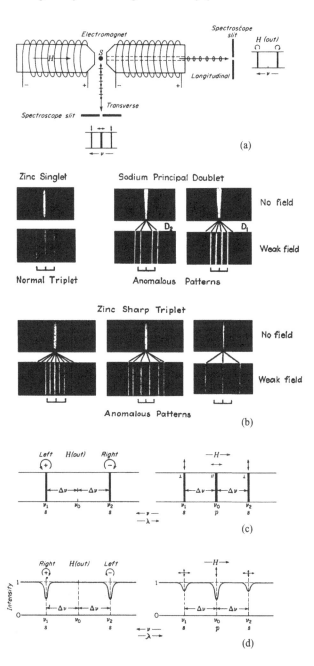

Fig. 3.1 The Zeeman effect, adapted from Jenkins and White (1950); reproduced with permission of McGraw-Hill Inc., New York. (a) Laboratory experiment to demonstrate the effect. (b) Zeeman-split spectral lines. (c) Zeeman patterns for a *normal triplet*, with polarization indicated. (d) A *normal triplet* in absorption.

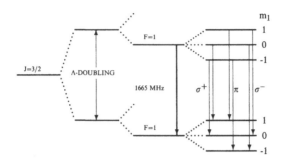

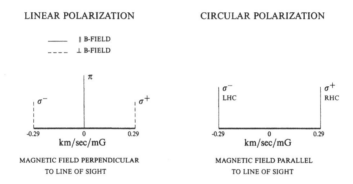

Fig. 3.2 The Zeeman effect for the OH spectral line at 1665 MHz after García-Barreto *et al.* (1988).

The Sun and similar stars

Photospheric fields in magnetically active regions are determined from the Zeeman effect (figs. 3.1 and 3.2; more details in Landstreet (1992)) and Hanle effect (Leroy 1985, Stenflo 1994); for the domains of applicability of such methods, see Landi Degl'Innocenti (1992), particularly the very enlightening discussion leading up to his figure 11. Three-dimensional field patterns as a function of time can be derived from maps of circular and linear polarization profiles of spectral lines, and these can be combined with Doppler data from the same specialized instrument. For a review of how much is known about solar magnetic field structure, see Stenflo (1989). (Figure 4 of that paper is a magnificent illustration of the Zeeman effect in a solar context, and some of the methods mentioned in that review can be applied to other stars). A recent conference volume (Schüssler and Schmidt 1994) gives a good impression of the detailed knowledge we have of solar magnetic fields, almost all of which is, in the end, based on the interpretation of polarimetric observations. For an impression of the sophistication of solar polarimetry at optical wavelengths, see Hagyard (1985), Keller and von der Lühe (1992), Stenflo (1994) and Volkmer (1994). The technique of Doppler imaging applied to Zeeman profiles can provide details of magnetic structure in chromospherically active stars (Donati *et al.* 1992). Coronal magnetic fields from solar flare regions are routinely detected by radio-polarimetry; the polarization is mostly circular. Attempts to detect linear X-ray polarization from these regions have yielded upper limits, which help to distinguish between models for the X-ray-generating mechanisms of flares. Flare stars and certain interacting binary stars with strong fields show radio polarization, probably by mechanisms similar to those in solar flares. For an overview of solar and stellar radio-polarimetry and the radiation mechanisms involved, see Dulk (1985) from which our table 3.1 has been adapted.

Magnetic A_p stars

The *longitudinal Zeeman effect* has been used extensively to analyse the strong magnetic fields which some of these rather exceptional stars have in large 'spots' on their surface. The apparent field varies with time, due to the sub-Earth point scanning the magnetic structure which is not aligned with the rotation axis. Magnetism in non-degenerate stars is reviewed in Borra *et al.* (1982) and Landstreet (1992). In the *transverse Zeeman effect*, the central component is twice the equivalent width of the outer ones. If an absorption line is strong, the difference in saturation will lead to a net linear polarization for the complete Zeeman multiplet. In a spectral region with many saturated lines, this will lead to a small net broad-band linear polarization (see Landstreet (1992, p. 43) for references).

Magnetic white dwarfs

Some white dwarfs have been found to have magnetic fields so strong that the Zeeman effect can be detected even in very wide features in the optical spectrum. For the stronger fields (up to hundreds of mega-gauss), magnetic effects completely distort the spectrum and produce both circular and linear polarization in lines and continuum. Fitting models to full spectro-polarimetry is the best attack in trying to understand these objects. Recent reviews are by Chanmugam (1992) and by Landstreet (1992); they list over 25 magnetic white dwarfs, usually detected by polarimetry of one kind or another and always confirmed by that technique. A recent study of a high-field white dwarf is Liebert *et al.* (1994), a low-field (1 MG!) example is discussed in Schmidt *et al.* (1992a). Multi-band time-resolved polarimetry is essential to the study of the

Table 3.1 *Polarized radio emission from the Sun*
Extracted from Table 1 of Dulk (1985).

Burst type	Duration	Polarization	Frequency range	Source
I	< 1 s	50–100%	50–300 MHz	large sunspots
II	> 10 min	low	200→1 MHz	flare shock wave
III	few seconds	< 30%	200→1 MHz	0.1–0.5 c electron stream
IV moving	≈30 min	low→high	0.2–1 GHz	small flare
IV flare continuum	≈20 min	0–40%	0.2–1 GHz	moderate to large flare
IV storm continuum	few hours	60–100%	300→50 MHz	flare, late phase
V	> 1 min	<10%, changes sign	100→10 MHz	follows some IIIs
microwave impulsive	> 1 min	≈30%	3–30 GHz	small to large flares
microwave IV	≈10 min	≈10%	1–30 GHz	large flares/shocks
microwave postburst	minutes–hours	low	1–10 GHz	flare, late phase
microwave spike burst	≈10 ms	≈100%	0.5–5 GHz	flare/hard X-rays

Arrows denote progression with time, over the frequency or polarization range.

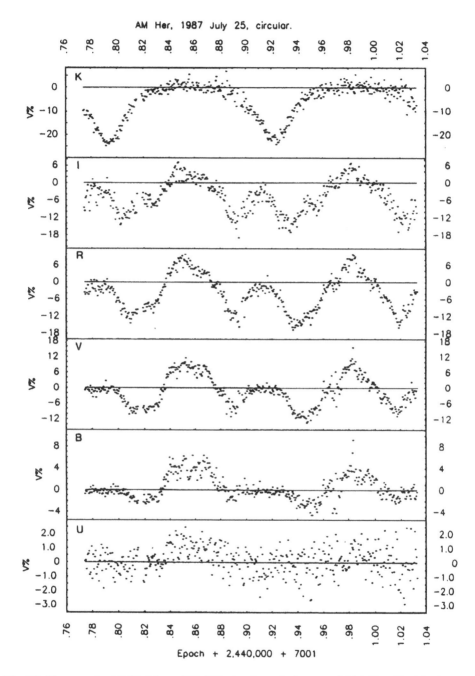

Fig. 3.3 Time-resolved 0.35–2.2 μm ('U'–'K') circular polarimetry of AM Her. 'V%' is the degree of circular polarization V/I, expressed as a percentage; from Hough *et al.* (1991). The polarimeter shown in fig. 6.4 was used to obtain these results.

internal structure of close binaries with a magnetic white dwarf component; fig. 3.3 shows such data for AM Her (Hough *et al.* 1991).

Pulsars

The magnetic field configuration and radiation mechanism of pulsars have been studied in great detail by measuring the polarization variations during the radio pulses (average and individual, linear and circular polarization). Some details are given in the pulsar review by Taylor and Stinebring (1986); see Stinebring *et al.* (1984) for a full account of the polarimetry (see also fig. 3.12). Detailed optical polarimetry of the pulsar in the Crab Nebula is reported in Smith *et al.* (1988). Chanmugam (1992) reviews the available data on magnetic fields of neutron stars. An impression of the surprisingly detailed knowledge of pulsars obtained from polarimetry may be gained from Michel (1991).

Galactic magnetic fields (interstellar, transverse component)

These can be detected by radio (synchrotron emission) or optical (polarized extinction) methods. Both our own Galaxy and nearby spiral galaxies have been analysed (fig. 3.4). For reviews see Beck (1993), Wielebinski and Krause (1993) and Kronberg (1994). In the radio domain, 'Faraday rotation' of the plane of polarization (section 3.3.2) is used to estimate the strength of the longitudinal field component in the region between source and observer; the sources used range from local Galactic synchrotron emission (Brouw and Spoelstra 1976) and pulsars (Rand and Kulkarni 1989) to remote radio galaxies and quasars (fig. 3.5). Zeeman measurements of the longitudinal component are available for neutral hydrogen clouds in our Galaxy, seen in *absorption* against strong point sources of continuum radiation (Schwarz *et al.* 1986). Under suitable circumstances, magnetic fields in 21-cm *emission* regions can also be measured (Troland and Heiles (1982a,b); however, see the discussion in section 5.5.5).

Molecular clouds

Zeeman measurements of the longitudinal magnetic field as a function of gas density within molecular clouds are reviewed by Myers and Goodman (1988).

Masers

The longitudinal magnetic field within H_2O maser regions has been measured by Fiebig and Güsten (1989), and within an OH maser by García-Barreto *et al.* (1988); the latter used the very sophisticated technique of polarization-spectral-VLBI.

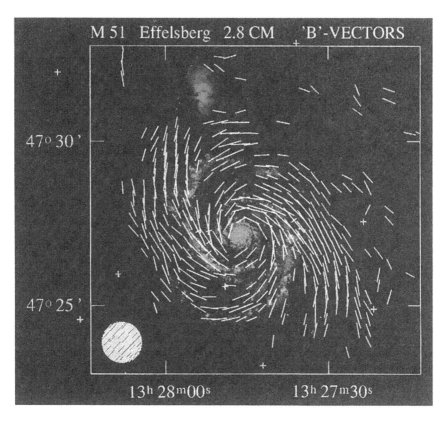

Fig. 3.4 Large-scale magnetic field structure of M51, as derived from 2.8 cm polarimetry with the Effelsberg radio telescope (Neininger (1992), optical image from Lick Observatory). The orientation of the polarization lines corresponds to the magnetic field orientation, their length is proportional to *polarized intensity*; this is a representation often used for radio-polarimetry of (presumed) synchrotron emission. Radio-polarimetry of galaxies does not suffer from contamination by scattered light, which complicates the interpretation of optical polarimetry of such objects; however, the wavelength of observation must be short enough for elimination of Faraday rotation to be possible.

Supernova remnants

The synchrotron emission from the Crab Nebula has been studied by polarimetry throughout most of the spectrum; see fig. 3.6 for an example. Numerous other supernova remnants have been studied in the radio domain, using polarimetry to obtain the magnetic field configuration (see *Astronomy and Astrophysics Abstracts*).

Quasars and active galaxies

Magnetic fields in these objects have been investigated by radio-polarimetry with increasing resolution, first by using single telescopes, later aperture synthesis instruments and most recently by VLBI networks. The polarization is

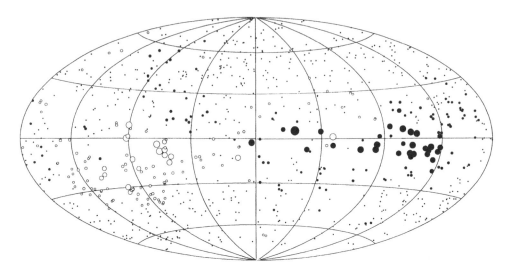

Fig. 3.5 Galactic magnetic field topology as revealed by observed rotation measure (RM) of extragalactic sources (RM = Faraday rotation normalized to a wavelength of 1 m). The sign of the longitudinal magnetic field component (filled vs open circles) is the most reliable datum from such a plot; to interpret the magnitude of the RM (size of the circles) in terms of magnetic field strength, additional data are needed (e.g. the electron density, the thickness of the Galactic gas disk and the scale of the irregularities of the Galactic field). This provisional map, showing large-scale Galactic magnetic field structure, was kindly provided by P.P. Kronberg, and displays the average rotation measure at the position of each of 901 sources (RM averaged over several neighbours). A similar plot without averaging, also due to Kronberg, is shown by Wielebinski and Krause (1993, figure 3), and is more suitable for tracing local anomalies in the magnetic field structure.

almost entirely linear (transverse field) and arises from synchrotron emisssion; Faraday rotation within the source is used where possible to estimate the longitudinal field. For general reviews see Saikia and Salter (1988) and Kronberg (1994), for the jets Bridle and Perley (1984). Polarimetry at optical and X-ray wavelengths can be used to investigate the central engine (which is often unresolved) of such sources (e.g. Mészáros *et al.* 1988). Optical polarization angles are also combined with radio morphology in attempts to unravel the mechanism of double-lobe formation.

3.2 Scattering geometry as a source of polarization

There are a number of situations in astronomy in which scattered radiation reaches the observer. The direction of vibration of the electric vector of the scattered radiation is at right angles to the *scattering plane*, the plane containing the incident and the scattered rays. Often direct radiation is also present; in some cases it is absent. Measurement of the linear polarization can help to

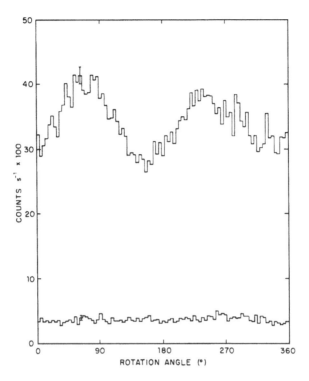

Fig. 3.6 X-ray linear polarimetry of the Crab Nebula, from Weisskopf *et al.* (1978). The curves show the variation of the detector signal with position angle of the instrument. The upper curve is the source, the lower represents the background signal to be subtracted. The wavelength was 4.77 Å (2.6 keV).

identify the scattering mechanism, it can pinpoint an obscured source, or it can give information on the properties of source (e.g. orientation as projected on the sky, spottedness) and/or the scattering medium (e.g. size, shape, degree of alignment and refractive index of the particles).

Figs. 3.7 and 3.8 illustrate applications of such techniques in the infrared; for more detail, see Aspin *et al.* (1990) and Minchin *et al.* (1991). The equivalent at optical wavelengths, as carried out with the polarimeter shown in fig. 6.6, is summarized in Scarrott (1991). Proposed applications at X-ray wavelengths may be found in Matt *et al.* (1989) and in Kaaret *et al.* (1989); these concern X-ray emission from such sources as Seyfert nuclei and black hole environments, subsequently scattered by an accretion disc. The extensive body of optical polarimetry of planets is reviewed in Coffeen and Hansen (1974), and for Venus in particular in Van de Hulst (1980, section 18.1.5). Linear polarimetry (optical, infrared, spectro-) promises to constrain very effectively the models for stars evolving from the asymptotic giant branch to planetary nebulae (Johnson and Jones 1991); Trammell *et al.* (1994) present spectro-polarimetry of post-AGB

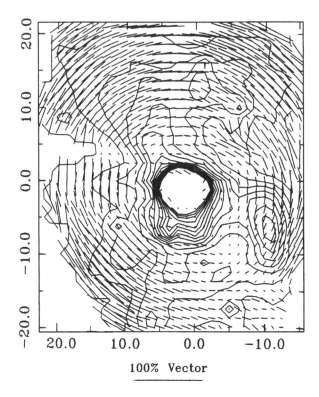

100% Vector

Fig. 3.7 A perfect example of scattering polarization (from Aspin *et al.* 1990). Linear
polarimetry, at a wavelength of 2.2 μm, of the region around the infrared source IRS4;
coordinate units are seconds of arc, IRS4 is at (0,0). The centro-symmetric arrangement
of the polarization lines persists out to at least 3 arcmin, showing that the central
source illuminates the whole of the region around it.

stars as convincing evidence that non-spherical structure is already present in
this early phase of planetary nebula formation.

   In general, the scattered photon may be at the same frequency as the
incident photon (as in Rayleigh and Thomson scattering) or the frequency
may be different (as in Compton and Raman scattering). *Spectro*-polarimetry
can be particularly useful in separating different components of the radiation
(optical applications in figs. 3.9 and 3.10).

3.3 Polarization dependence of the refractive index

If the refractive index of the medium between the source and the observer is
a function of polarization, radiation from the source will have its polarization
modified by this intervening medium, and this produces a number of observable effects. In terms of the complex notation for electromagnetic waves (see

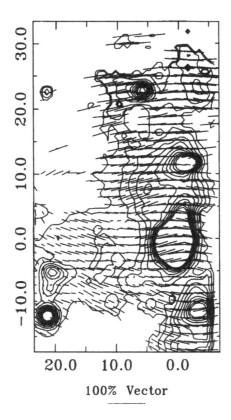

100% Vector

Fig. 3.8 Scattered radiation from a hidden source, from Walther *et al.* (1990). Linear polarimetry at 2.2 μm of the region near the infrared source IRS1. Coordinate units are seconds of arc, IRS1 is at (0,0). The authors comment: '...clearly shows centro-symmetric vectors about a point in the dark lane, indicating that this is the location of the exciting star. The polarization is extremely high, indicating that the dust must be optically thin to scattering. The geometry must be such that scattering takes place preferentially off large dust grains in the walls of the outflow...'.

section 4.2), the 'complex refractive index' determines the optical path length (real part; classical refractive index) and the attenuation (imaginary part; extinction coefficient) as the wave propagates. When these components of the complex refractive index depend on the polarization of the wave, one uses the terms birefringence (real part) and dichroism (imaginary part) for the difference between the values for two orthogonal polarization forms; these polarization forms may be linear, circular or elliptical, depending on the eigenmodes of the medium (see chapter 2). Examples are linear dichroism of a sheet polarizer and elliptical birefringence of a magnetized plasma. The following subsections describe astronomical occurrences of polarization dependence of the refractive index. Note that the concept of complex refractive index is a macroscopic description; the microscopic mechanisms underlying the polarization dependence

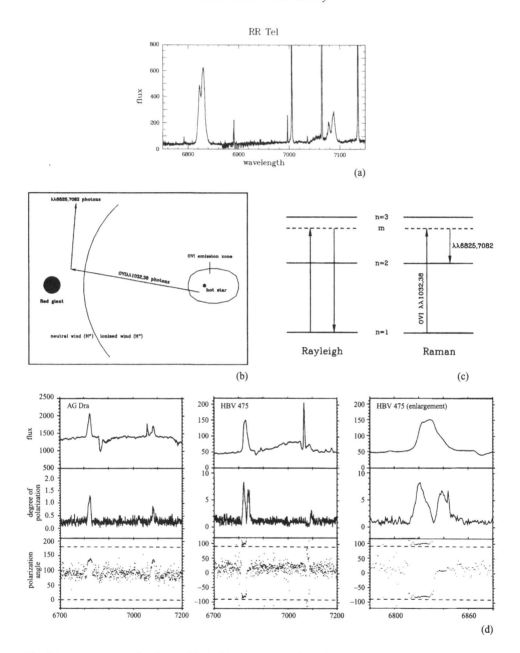

Fig. 3.9 Raman scattering in symbiotic binary stars. Adapted from Schild and Schmid (1992), where more detail is given; (a) is from Duerbeck and Schwarz (1995). The broad emission features in the flux spectra are due to Raman scattering (see energy level diagram (c)) of ultraviolet photons by the atoms of the neutral wind from the red giant (see plan view of the binary star (b)). Fractional Doppler width is increased in the ratio of the scattered to incident wavelengths. Part (d) shows spectro-polarimetry of two such stars. The scattering origin of the lines is confirmed by this spectro-polarimetry and the geometry of this kind of object is thus open to investigation.

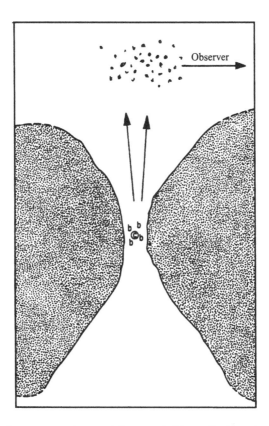

Fig. 3.10 A Seyfert galaxy core region model, suggested by optical spectro-polarimetry, adapted from Antonucci and Miller (1985). The spectrum of the central obscured source can be detected only through light scattered by electrons on the polar axis of the system. Such light will be linearly polarized, and spectro-polarimetry can be used to distinguish it from the other components of the radiation. Polarimetric observations play an important role in unified models for these remote extragalactic sources. See also Fosbury *et al.* (1993).

are varied, although magnetic fields play a role in most astronomical cases. (An exception to this last statement is the multi-wavelength polarization-sensitive radar reflectometry reported by Evans *et al.* (1994); this technique, similar to what is called 'ellipsometry' in optical laboratory practice, is very powerful for mapping planetary surface structures.)

Note: Like 'polarization' itself, the term 'dichroism' is a misnomer. It arose from the first known instance of this effect, the crystal K_2PdCl_4, in which the polarization dependence of the absorption coefficient is also wavelength-dependent, so that the colour of the crystal depended on the polarization (and the direction) of the light passing through it (see Clarke and Grainger 1971, p. 86). By analogy with 'birefringence', a term like 'biattenuance' would have been preferable to 'dichroism'.

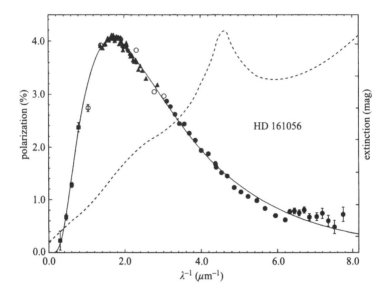

Fig. 3.11 Linear dichroism of the interstellar medium in the direction of the star HD 161056.
Starlight is polarized by dust particles partially aligned in an interstellar magnetic field. The
shape of the curve (solid line, from Somerville *et al.* (1994)) is more or less the same for
all stars ('Serkowski curve') when the degree of polarization and the wavelength are both
normalized at the peak of the curve. For reference, the extinction curve for this star has also
been sketched in (broken line, arbitrary zeropoint). At the wavelength of the polarization peak,
the difference in extinction for the two orthogonal polarizations is about equal to the thickness
of the dashed curve; this highlights the precision that must often be achieved in polarimetry.

3.3.1 Dichroism

Dichroism is the differential extinction of orthogonally polarized radiation
components. A 'dichroic' sheet of Polaroid operates by differential absorption
of linear polarization components, and something similar can happen at optical
wavelengths in interstellar space. Interstellar dust particles are non-spherical
and/or have crystalline structure; they have a different scattering cross-section
(effective area) for light linearly polarized parallel to the geometric or crystalline
axis than for light polarized at right angles to it. Since the non-sphericity of the
particles and the crystalline structure also influence the magnetic and electrical
properties, the interstellar magnetic field can induce a slight preferential orientation with respect to the field. These two deviations from perfect isotropy
cooperate to produce a certain amount of polarization by differential extinction
of the linear polarizations along and across the transverse magnetic field component (*extinction* denotes absorption plus scattering; at optical wavelengths it
is scattering that matters, further into the infrared absorption becomes more
important; polarized true absorption will – according to Kirchhoff's law – be
accompanied by polarized emission; see Aitken *et al.* (1986) and Hildebrand

(1988)). The degree of polarization of light from distant stars peaks (fig. 3.11) at a wavelength which is related to the median size of the interstellar grains (Serkowski *et al.* (1975); see Martin and Whittet (1990), Whittet *et al.* (1992) and Somerville *et al.* (1994) for the latest information and extensive references). Observations of interstellar polarization are used to investigate both the properties of the grains and the magnetic field topology.

3.3.2 Birefringence

The other way in which the intervening medium can influence the state of polarization is by birefringence of the medium. For birefringence to have any effect, the radiation generally must have been polarized elsewhere in the first place (but see section 7.1).

Linear birefringence is expected to occur at optical wavelengths in the interstellar medium, due to the optical properties of the aligned dust particles that also cause the linear polarization. This linear birefringence can convert previously generated linear polarization into circular polarization, if the magnetic field is twisted in some systematic way. The interstellar circular polarization is very small, and good measurements are only available for a few stars; the main use of this phenomenon has been for estimating the (real part of the) refractive index of the interstellar grains. Deguchi and Watson (1985) have estimated linear birefringence effects in absorption lines at radio wavelengths; in this case, the birefringence is due to unequal populations in the magnetic substates of atoms and molecules.

Circular birefringence due to a longitudinal magnetic field component is called *Faraday rotation* and is an important observable at radio wavelengths. Faraday rotation is proportional to the square of the wavelength, which means that there is only a relatively small wavelength range over which it is observable in any one application: if the wavelength is too short, the rotation is too small to detect; if it is too long, the inhomogeneities in the intervening medium cause many different values of the rotation to be present within the telescope beam and no net polarization remains (Stokes Q and U average out to zero). On the other hand, the wavelength-squared dependence of polarization angle (in simple cases) is the one sure proof that we are measuring a linearly polarized component rather than some instrumental effect. It is also the only observable connected with the longitudinal field component when linear polarization is all we can detect (see Spoelstra (1984) and Sofue *et al.* (1986)): the quantity measured is the product of longitudinal magnetic field component and density of thermal electrons, integrated over the line of sight. Faraday rotation has

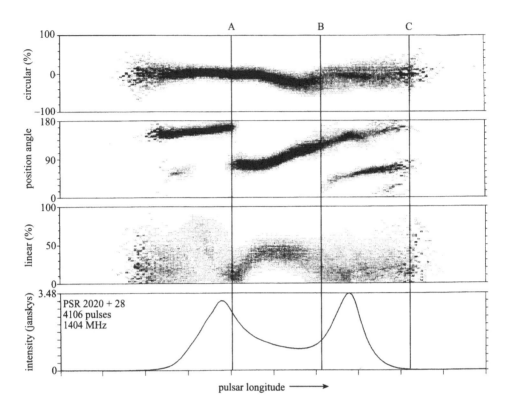

Fig. 3.12 Elliptically polarized eigenmodes in pulsar radiation, adapted from Stinebring *et al.* (1984). Polarization variations within the pulse of PSR2020+28 are shown. The average pulse is shown, together with a statistical display of the polarization of individual pulses. At A, the degree of linear polarization passes through 0, but the 90° jump in the polarization angle shows that for the locus in the (p_Q, p_U) diagram the origin (0,0) is nothing special. Between B and C, two orthogonal elliptical modes are possible, as evidenced by the two branches of linear polarization separated in angle by 90° and the spread of the circular component. The jumps from one linear branch to the other occur at the same moment as the sign reversals of the circular component.

been exploited to investigate fields within distant radio sources, the intergalactic medium and the Galaxy (see figs. 3.5 and 7.1).

For an arbitrary line of sight, a magnetized plasma will have elliptically polarized eigenmodes and will be *elliptically birefringent*. In general, elliptical birefringence will cause conversion from one form of elliptical polarization to another (conversions between Q, U and V). In astronomy, there is usually too little information to make it worthwhile to refer to the general case, and *quasi-longitudinal* or *quasi-transverse* conditions are invoked. There is evidence in the mode-switching of pulsars (Taylor and Stinebring 1986, p. 309) that elliptical eigenmodes are important in the mechanism of generating pulsar radio waves (fig. 3.12). Other cases of importance are generation of radio waves in stellar coronae and propagation of radio waves through the Earth's ionosphere.

4

Polarization algebra and graphical methods

This chapter introduces the tools used by astronomers and instrument designers in describing the action of a medium on the polarization of the radiation passing through it. In the majority of situations encountered in astronomy, the phase of the wave is unimportant, and we need a way to describe the transformation of Stokes parameters, i.e. the changing polarization forms which support the *flow of radiant energy*. For cases where the phase of the polarized radiation *is* important (e.g. polarization effects within an optical interferometer, the focusing of a plane wave by a radio telescope, or the amplification of polarized radiation within an astronomical maser), an alternative formulation will be introduced (in section 4.2) that describes the transformation (including phase) of the electric field vibrations of two orthogonal 100% polarized waves (usually linear polarization). In this formulation, partial polarization cannot be handled, and we must make separate calculations for two orthogonal polarizations of the incident radiation, constructing the incoherent sum at the end. Shurcliff (1962, sections 8.6, 8.7, 8.9) details the early history of these two calculi and compares their fields of use; a concise statement of the relationship between the two calculi may be found in Stenflo (1994, section 2.6).

4.1 Mueller matrices

As discussed in chapter 2, the four Stokes parameters denote the flow of radiant energy in specific vibrations of the electromagnetic field, and all four are expressed in the same units. We write them collectively as a four-element column matrix, which is generally referred to as the *Stokes vector*; as we have seen in sections 2.2 and 2.5, its elements are not all independent, so as a 4-vector it is somewhat limited. When convenient, the vector will be written as a row matrix, but a column matrix is always intended.

When radiation propagates through a certain volume of space (which may be empty or contain some material medium), the polarization of the input and output radiation is represented by the input and output Stokes parameters. Within the volume, the state of polarization may be altered: in general, any elliptical polarization may be *transformed* into some other elliptical polarization form, or the radiation may be *polarized* by the medium. This can be represented by a transformation between the input and output Stokes parameters, and in general the transformation is linear. When the input and output Stokes parameters are arranged as 4-vectors, the transformation becomes a 4×4 matrix $\mathbf{M}$:

$$\mathbf{S}_{\text{out}} = \mathbf{M} \cdot \mathbf{S}_{\text{in}}$$

where

$$\mathbf{M} = \begin{pmatrix} m_{11} & m_{12} & m_{13} & m_{14} \\ m_{21} & m_{22} & m_{23} & m_{24} \\ m_{31} & m_{32} & m_{33} & m_{34} \\ m_{41} & m_{42} & m_{43} & m_{44} \end{pmatrix}$$

and $\mathbf{S}$ stands for the Stokes vector (I, Q, U, V). Since Stokes parameters are real quantities, the elements of $\mathbf{M}$ are all real numbers; m_{11} must be positive (I is always positive), and the other elements can be positive or negative. The matrices are known as *Mueller matrices*, after H. Mueller who worked out their precise form for a number of optical components (Mueller 1948); see Shurcliff (1962, pp. 117–18) for the historical development of these ideas. The assumption of linear transformations amounts to assuming that there is no functional dependence of the elements of the matrix on the incident radiation (e.g. no processes such as squaring the amplitude in a mixer, frequency doubling, etc.; frequency conversion – though involving a mixer stage – can, as a single indivisible operation, be considered a linear transformation, see section 6.2.1).

When the radiation travels through several media in succession, the output Stokes vector for one medium ('a') is the input Stokes vector for the next ('b'):

$$\mathbf{S}_{\text{b,out}} = \mathbf{M}_{\text{b}} \cdot \mathbf{S}_{\text{b,in}} \equiv \mathbf{M}_{\text{b}} \cdot \mathbf{S}_{\text{a,out}} = \mathbf{M}_{\text{b}} \cdot \mathbf{M}_{\text{a}} \cdot \mathbf{S}_{\text{a,in}} \equiv \mathbf{M} \cdot \mathbf{S}_{\text{a,in}}$$

or

$$\mathbf{M} = \mathbf{M}_{\text{b}} \cdot \mathbf{M}_{\text{a}}$$

where $\mathbf{M}$ represents the combined action of the two media 'a' and 'b'; it is the matrix product of $\mathbf{M}_{\text{b}}$ and $\mathbf{M}_{\text{a}}$ (note the order: the first medium traversed comes last in the matrix equation). This procedure is used extensively in the design of 'optical' instruments (see chapter 6) and in the representation of the transformations of polarized radiation within a multiple or distributed astronomical source of polarized radiation (e.g. a stellar or planetary atmosphere,

Table 4.1 *Simple Mueller matrices*

See Shurcliff (1962) and Kliger *et al.* (1990) for more complicated examples.

Polarizers

linear $(+Q)$ polarizer	linear $(-Q)$ polarizer	linear $(+U)$ polarizer
$k\begin{pmatrix} 1 & 1 & 0 & 0 \\ 1 & 1 & 0 & 0 \\ 0 & 0 & 0 & 0 \\ 0 & 0 & 0 & 0 \end{pmatrix}$	$k\begin{pmatrix} 1 & -1 & 0 & 0 \\ -1 & 1 & 0 & 0 \\ 0 & 0 & 0 & 0 \\ 0 & 0 & 0 & 0 \end{pmatrix}$	$k\begin{pmatrix} 1 & 0 & 1 & 0 \\ 0 & 0 & 0 & 0 \\ 1 & 0 & 1 & 0 \\ 0 & 0 & 0 & 0 \end{pmatrix}$

homogeneous circular $(+V)$ polarizer	homogeneous circular $(-V)$ polarizer	linear $(+Q)$ polarizer + quarterwave $(\eta = 45^\circ)$
$k\begin{pmatrix} 1 & 0 & 0 & 1 \\ 0 & 0 & 0 & 0 \\ 0 & 0 & 0 & 0 \\ 1 & 0 & 0 & 1 \end{pmatrix}$	$k\begin{pmatrix} 1 & 0 & 0 & -1 \\ 0 & 0 & 0 & 0 \\ 0 & 0 & 0 & 0 \\ -1 & 0 & 0 & 1 \end{pmatrix}$	$k\begin{pmatrix} 1 & 1 & 0 & 0 \\ 0 & 0 & 0 & 0 \\ 0 & 0 & 0 & 0 \\ 1 & 1 & 0 & 0 \end{pmatrix}$

Retarders

linear retarder halfwave $(\eta = 0^\circ$ or $90^\circ)$	linear retarder halfwave $(\eta = \pm 45^\circ)$
$k\begin{pmatrix} 1 & 0 & 0 & 0 \\ 0 & 1 & 0 & 0 \\ 0 & 0 & -1 & 0 \\ 0 & 0 & 0 & -1 \end{pmatrix}$	$k\begin{pmatrix} 1 & 0 & 0 & 0 \\ 0 & -1 & 0 & 0 \\ 0 & 0 & 1 & 0 \\ 0 & 0 & 0 & -1 \end{pmatrix}$

linear retarder quarterwave $(\eta = 0^\circ)$	linear retarder quarterwave $(\eta = 90^\circ)$	linear retarder quarterwave $(\eta = \pm 45^\circ)$
$k\begin{pmatrix} 1 & 0 & 0 & 0 \\ 0 & 1 & 0 & 0 \\ 0 & 0 & 0 & 1 \\ 0 & 0 & -1 & 0 \end{pmatrix}$	$k\begin{pmatrix} 1 & 0 & 0 & 0 \\ 0 & 1 & 0 & 0 \\ 0 & 0 & 0 & -1 \\ 0 & 0 & 1 & 0 \end{pmatrix}$	$k\begin{pmatrix} 1 & 0 & 0 & 0 \\ 0 & 0 & 0 & \mp 1 \\ 0 & 0 & 1 & 0 \\ 0 & \pm 1 & 0 & 0 \end{pmatrix}$

Various

isotropic absorber	ideal depolarizer	rotation by $+\theta$
$k\begin{pmatrix} 1 & 0 & 0 & 0 \\ 0 & 1 & 0 & 0 \\ 0 & 0 & 1 & 0 \\ 0 & 0 & 0 & 1 \end{pmatrix}$	$k\begin{pmatrix} 1 & 0 & 0 & 0 \\ 0 & 0 & 0 & 0 \\ 0 & 0 & 0 & 0 \\ 0 & 0 & 0 & 0 \end{pmatrix}$	$\begin{pmatrix} 1 & 0 & 0 & 0 \\ 0 & \cos 2\theta & \sin 2\theta & 0 \\ 0 & -\sin 2\theta & \cos 2\theta & 0 \\ 0 & 0 & 0 & 1 \end{pmatrix}$

k is the transmittance $(I_{\text{out}}/I_{\text{in}})$ for unpolarized light and is ≤ 0.5 for the polarizers, ≤ 1 for the other components; η is the orientation of the component ($\eta = 0^\circ$ denoting principal orientation).

Faraday rotation within a synchrotron source). In the radio domain, a similar but distinct development has occurred: 4×4 matrices of a slightly different kind are used to describe the polarization response of a correlation-type interferometer (Hamaker *et al.* 1995).

For many purposes (such as reading this book), multiplication is the only matrix algebra one needs. It is specified by

$$m_{ij} = \sum_{k=1}^{4} (m_{\mathrm{b}})_{ik} \cdot (m_{\mathrm{a}})_{kj}$$

or, as a pictogram:

$$m_{23} = \sum_{k=1}^{4} \begin{pmatrix} \cdot & \cdot & \cdot & \cdot \\ \rightarrow & \rightarrow & \rightarrow & \rightarrow \\ \cdot & \cdot & \cdot & \cdot \\ \cdot & \cdot & \cdot & \cdot \end{pmatrix} \times \begin{pmatrix} \cdot & \cdot & \downarrow & \cdot \\ \cdot & \cdot & \downarrow & \cdot \\ \cdot & \cdot & \downarrow & \cdot \\ \cdot & \cdot & \downarrow & \cdot \end{pmatrix}$$

Modern papers in both astrophysics and instrumentation increasingly use matrix methods of some sophistication (e.g. Sánchez Almeida and Martínez Pillet 1992, McClain *et al.* 1993, Hovenier 1994, Stenflo 1994, Hamaker *et al.* 1995); more insight into matrix algebra and the structure of Mueller matrices will certainly pay off in future work. The Mueller matrices of a number of components were derived long ago and are tabulated in, for instance, Shurcliff (1962) or Kliger *et al.* (1990). A few worked examples will suffice here, to demonstrate the simple-minded phenomenological approach one may often use; table 4.1 lists a few more. The derivations are not always intuitive, mainly because we do not have a clear mental picture of the Stokes vector and its properties. It is worth investing some time understanding the matrices in this chapter, including those of fig. 4.1 and table 4.2; programs like Mathematica simplify matrix manipulation considerably, but without some basic understanding such manipulations can be dangerous.

Example 1: An ideal linear polarizer absorbs all the light of one linear polarization ($-Q$, say) and passes all the light of the opposite linear polarization ($+Q$). Therefore, if the input is unpolarized light of intensity I, the output intensity is $0.5I$ and, since the output light is fully polarized, Q must also equal $0.5I$, while U and V are zero. Hence, for a $+Q$ polarizer:

$$(0.5I, 0.5I, 0, 0) = \mathbf{M} \cdot (I, 0, 0, 0)$$

This implies that m_{11} and m_{21} are both 0.5, while m_{31} and m_{41} must be zero; it does not tell us anything about the other elements. We now pass fully circularly polarized light through the linear polarizer; since this can be regarded as being made up of two orthogonal linear polarizations, again half the intensity will be passed, in the form of linear polarization $(+Q)$:

$$(0.5I, 0.5I, 0, 0) = \mathbf{M} \cdot (I, 0, 0, I)$$

Given the previous m_{*1} elements, this can only be true if m_{14} and m_{24} are zero; for the output U and V to be zero, we also need m_{34} and m_{44} equal to zero. A similar argument with U substituting for V leads to m_{*3} being zero. We thus have as the Mueller matrix of the ideal linear $+Q$ polarizer:

$$\begin{pmatrix} 0.5 & . & 0 & 0 \\ 0.5 & . & 0 & 0 \\ 0 & . & 0 & 0 \\ 0 & . & 0 & 0 \end{pmatrix}$$

We now pass 100% linearly polarized light of the correct orientation to the polarizer; since the light already has the correct polarization, all of it must pass through unchanged, so

$$(I, I, 0, 0) = \mathbf{M} \cdot (I, I, 0, 0)$$

This can only be true if m_{12} and m_{22} are 0.5 and if m_{32} and m_{42} are zero. The matrix is generally written as:

$$\mathbf{M}_{\text{lin.pol.}}(0°) = k \cdot \begin{pmatrix} 1 & 1 & 0 & 0 \\ 1 & 1 & 0 & 0 \\ 0 & 0 & 0 & 0 \\ 0 & 0 & 0 & 0 \end{pmatrix}$$

where k is the transmittance of the polarizer for unpolarized light (0.5 for a perfectly transparent polarizer and close to that for many crystal polarizers shown in fig. 6.3; for Polaroids, k is between 0.2 and 0.4, depending on wavelength).

Example 2: What happens when we rotate the polarizer through 90°, or consider the second beam of a two-beam crystal or wire-grid polarizer? The output vector will have its Q reversed: $(0.5I, -0.5I, 0, 0)$. The arguments concerning unpolarized and circularly polarized input remain the same, except that Q_{out} must equal $-0.5I$, implying that m_{21} must be -0.5. If we now take linear $-Q$ polarization as the input (Stokes vector: $I,-I,0,0$), all of the I should come through, and Q_{out} should be equal to $-I$; this means that m_{12} must be -0.5 and m_{22} must be 0.5:

$$\mathbf{M}_{\text{lin.pol.}}(90°) = k \cdot \begin{pmatrix} 1 & -1 & 0 & 0 \\ -1 & 1 & 0 & 0 \\ 0 & 0 & 0 & 0 \\ 0 & 0 & 0 & 0 \end{pmatrix}$$

Similar arguments hold for the polarizer at $\pm 45°$. Rather than going through these, let us look at the general case of rotation of components with respect to the coordinate system.

Example 3: If we rotate an optical component, its Mueller matrix will remain the same, as long as we express it in a coordinate system that rotates with the component. To express the effect of the rotated component in the original coordinate system, we must first transform the input Stokes vector into the rotated coordinate system, then apply the component matrix that we know, then transform back into the original coordinate system. Since rotating the coordinate system does not change the nature of the radiation, it must be just the Stokes parameters that are transformed into each other, so that rotation must also be expressible as a Mueller matrix. The matrix of a rotated component must therefore be:

$$\mathbf{M}_\theta = \mathbf{T}(-2\theta) \cdot \mathbf{M}_0 \cdot \mathbf{T}(2\theta)$$

where $\mathbf{T}(2\theta)$ is the rotator matrix describing the change of coordinate system. What form should $\mathbf{T}$ take? Rotation of the coordinate system should not involve I or V, as they do not contain the polarization angle χ. If we rotate the coordinate system through $+\theta$ (see fig. 2.1; θ in the same sense as χ),

$$\chi_{new} = \chi_{old} - \theta$$

$$Q_{new} = +Q_{old} \cdot \cos 2\theta + U_{old} \cdot \sin 2\theta$$

$$U_{new} = -Q_{old} \cdot \sin 2\theta + U_{old} \cdot \cos 2\theta$$

or

$$\mathbf{T}(2\theta) = \begin{pmatrix} 1 & 0 & 0 & 0 \\ 0 & \cos 2\theta & \sin 2\theta & 0 \\ 0 & -\sin 2\theta & \cos 2\theta & 0 \\ 0 & 0 & 0 & 1 \end{pmatrix}$$

Applying this, we find the Mueller matrix of the linear polarizer at orientation θ to be:

$$k \cdot \begin{pmatrix} 1 & \cos 2\theta & \sin 2\theta & 0 \\ \cos 2\theta & (\cos 2\theta)^2 & \cos 2\theta \cdot \sin 2\theta & 0 \\ \sin 2\theta & \cos 2\theta \cdot \sin 2\theta & (\sin 2\theta)^2 & 0 \\ 0 & 0 & 0 & 0 \end{pmatrix}$$

Example 4: When a medium is birefringent, it introduces a phase difference between its two eigenmodes. The most common form of birefringence used in instruments is linear birefringence. Slabs of such birefringent crystal are called halfwave or quarterwave plates, if the phase difference is exactly one-half or one-quarter of a wavelength; in general, they are called linear *retarders*, and the phase difference is called their *retardance*. A quarterwave plate oriented at 45° to the plane of electric vibration of input linearly polarized light transforms this polarization into circular polarization; it is the hardware analogue of the thought experiment by which we introduced circularly polarized light in chapter 2. Similarly, circularly polarized light incident on a quarterwave plate is transformed into linear polarization, with the plane of electric vibration at 45° to those of the eigenmodes of the crystal. By such arguments one can show that a quarterwave plate has the Mueller matrix

$$\begin{pmatrix} 1 & 0 & 0 & 0 \\ 0 & 1 & 0 & 0 \\ 0 & 0 & 0 & 1 \\ 0 & 0 & -1 & 0 \end{pmatrix}$$

This form is for orientation 0°, i.e. with the plane of electric vibration of the fast eigenmode (that with the lowest refractive index) along the reference zero direction of the coordinate system. Such orientation is also called *principal orientation*. It is easy to see that I and Q are not affected, but U (linear polarization along ±45°) and V (circular polarization) are transformed into each other.

It can be shown (by using the third form of the Stokes parameters given in section 4.3) that the Mueller matrix for a general linear retarder (with retardance Δ) in its principal orientation is

$$\begin{pmatrix} 1 & 0 & 0 & 0 \\ 0 & 1 & 0 & 0 \\ 0 & 0 & \cos \Delta & \sin \Delta \\ 0 & 0 & -\sin \Delta & \cos \Delta \end{pmatrix}$$

Stenflo (1994, p. 320) discusses the case of a partially polarizing retarder (e.g. reflection at an inclined metallic surface).

Example 5: 'Depolarizers' are components used in the optical part of the spectrum to calibrate equipment or to stabilize the response of a polarization-sensitive component. For complete depolarization of an arbitrarily polarized input signal, the Mueller matrix for a depolarizer must take the form:

$$\begin{pmatrix} d_{11} & d_{12} & d_{13} & d_{14} \\ 0 & 0 & 0 & 0 \\ 0 & 0 & 0 & 0 \\ 0 & 0 & 0 & 0 \end{pmatrix}$$

If a component existed with non-zero values of d_{12}, d_{13} or d_{14}, its transmission would depend on input polarization, i.e. it would be a polarization *analyser*. Although no analyser is known that is not at the same time a *polarizer* for unpolarized radiation (non-zero d_{21}, d_{31} or d_{41}), a rigorous proof that such a component cannot exist is as yet lacking.

No physical component corresponds exactly to the definition above; arrangements involving multiple scattering would probably come closest. All 'depolarizers' in use actually obey the definition:

$$\mathbf{D} \equiv \int \int \int \int \mathbf{M}(\lambda, t, A, \theta) \, \mathrm{d}\lambda \, \mathrm{d}t \, \mathrm{d}A \, \mathrm{d}\theta \approx \begin{pmatrix} d_{11} & 0 & 0 & 0 \\ 0 & 0 & 0 & 0 \\ 0 & 0 & 0 & 0 \\ 0 & 0 & 0 & 0 \end{pmatrix}$$

This expresses the aim that, *averaged* over the wavelength range, time, beam size, component position angle (or a combination of these, in other words some sort of ensemble average), the polarization of the input radiation should be destroyed. A true depolarizer must include randomness in the ensemble-averaging, but most practical 'depolarizers' are deterministic, so they should be referred to as 'pseudo-depolarizers' (the depolarization can be reversed by a suitably constructed piece of equipment or the polarized components in the mixture can be identified). Spatially almost-random averaging can be obtained by using a rough-surfaced wave plate cemented to a smooth cover plate (Peters 1964).

A Lyot pseudo-depolarizer integrates the action of two stationary multi-wave plates over wavelength and only works well for broad passbands with 'soft' edges. Halfwave and/or quarterwave plates rotating at constant speed can be used in narrow wavelength bands to integrate over the position angle, which in this case is a function of time. A *near-halfwave* plate rotating at constant speed, of retardation $\pi - b$ ($b \ll 1$ and a function of wavelength), has a Mueller matrix (averaged over an integral number of quarter-revolutions):

$$k \cdot \begin{pmatrix} 1 & 0 & 0 & 0 \\ 0 & l & 0 & 0 \\ 0 & 0 & l & 0 \\ 0 & 0 & 0 & -\cos b \end{pmatrix}$$

where $l = (1 - \cos b)/2 \approx b^2/4$ and k is the transmission for unpolarized light. This represents a *linear pseudo-depolarizer*, a component which, on average, reduces linear polarization very effectively, while inverting but only slightly reducing the circular component. Such an optical component is very useful for reducing photometric errors due to instrumental linear polarization effects, and also when one is trying to detect the minute interstellar circular polarization in the presence of linear polarization, which is stronger by about two orders of magnitude. The 'superachromatic' halfwave plate illustrated in fig. 6.7 performs very well in this application ($|b| < 0.05$ at all wavelengths within a very wide range). Effective use of random rotation of the entire telescope for depolarization of an unwanted linear polarization component is described in section 5.5.6.

A depolarizer is encountered in astrophysics in the form of 'Faraday de-polarization' (e.g. Beck 1993). In the quasi-longitudinal approximation of magneto-ionic theory, this is represented by the depolarizer matrix:

$$\mathbf{D_{Far}} = \int_{\Delta\lambda} \int_{\Omega_{beam}} \begin{pmatrix} 1 & 0 & 0 & 0 \\ 0 & \cos 2\theta_F & \sin 2\theta_F & 0 \\ 0 & -\sin 2\theta_F & \cos 2\theta_F & 0 \\ 0 & 0 & 0 & 1 \end{pmatrix} d\lambda \, d\Omega$$

where the Faraday rotation $\theta_F = \lambda^2 \cdot \xi(\Omega)$, $\Delta\lambda$ is the passband and Ω_{beam} is the beam solid angle. For $\theta_F \ll 1$ radian, the plane of polarization is rotated, but the degree of polarization is not affected. When $\theta_F > 1$ radian, however, discrete directions within the beam and discrete wavelengths within the passband can have widely different values of θ_F, and the result is *linear* depolarization of any radiation from behind the 'Faraday screen'. This screen itself may be distributed along the line of sight, possibly overlapping the source of polarized radiation; in such a case the integral becomes still more complicated. The pioneering paper on this subject, used extensively since, is Burn (1966). In a modern study, A.D. Poezd and A. Shukurov (Moscow State University, private communication, 1995) include the effects of the higher angular resolution of modern observations (a finite number of turbulent magnetic field cells within the beam); they identify ways of distinguishing between scatter in the intrinsic source polarization angle and a distribution of Faraday rotations along the line of sight. This astrophysical example illustrates how pseudo-depolarization results from the averaging of many mutually incoherent contributions of differing polarization; at any one wavelength and angular resolution, the apparent result is depolarization, but the amount of depolarization varies both with wavelength and with resolution, so that by suitable observations and reconstruction (modelling) one may identify individually polarized radiation components, i.e. pseudo-depolarization is the correct term.

> **Note:** In works translated from French, one may come across 'depolarization' in the sense of 'separation into two polarized components'. In English, this is confusing in the extreme.

Example 6: Graphics and pictograms of Mueller matrices can be very useful in summarizing the action of components (e.g. figs. 4.1 and 4.2, and table 4.2).[*]

Example 7: To simplify certain calculations, Collett (1993, p. 164 *et seq.)* and Stenflo (1994, p. 254) introduce so-called 'diagonalized Mueller matrices'. These are not Mueller matrices in the conventional sense; their elements are in general complex (which is the price one pays for diagonalizing). To convert them into conventional Mueller matrices, they must be pre- and post-multiplied by certain unitary matrices with complex elements; in other words, they belong to a representation in terms of four (complex) linear combinations of the Stokes parameters. Similarly, Hovenier and Van der Mee (1983, equations (19)–(48)) mention other linear combinations of the Stokes parameters, with corresponding 4 × 4 matrices (e.g. a 'CP' or circular polarization representation, in which the rotation matrix is diagonalized). To avoid confusion, none of the above matrices should really be called *Mueller* matrices. For more comment and references on what, in a particular application, may be the most useful representation, see Van de Hulst (1980, pp. 497–8).

In radio synthesis instruments, the 'Stokes visibilities' in the pupil or aperture plane are the spatial (complex) Fourier transforms of the (real) Stokes parameter sky distributions. Although these visibilities are complex, they are sometimes referred to as 'Stokes parameters', and 4 × 4 matrices transforming these quantities are considered to be 'complex Mueller matrices'. It should be noted that each of the Stokes visibilities, though complex, depends only on the one corresponding Stokes parameter sky distribution; the complex representation is used only for the Fourier transform relation, and the 'complex Mueller matrices' operate in the pupil domain.

---

[*] Note added in proof: see also J. L. Pezzaniti and R. A. Chipman (1995), *Opt. Eng.* **34**, 1558–68.

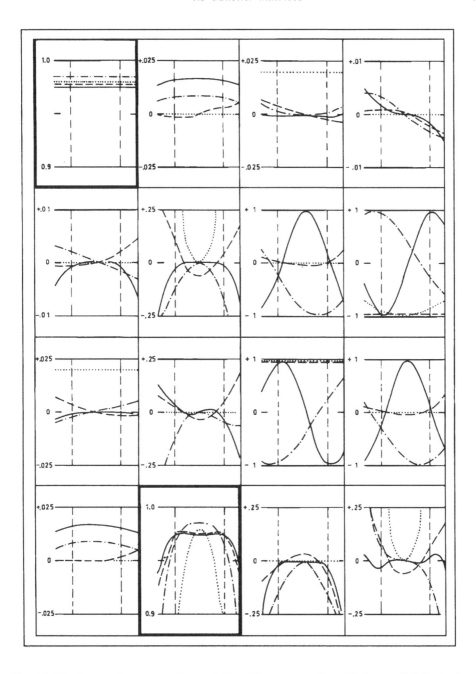

Fig. 4.1 Mueller matrix spectra of several $Q \rightarrow V$ converters, from Tinbergen (1973). Each element of the matrix is shown in graphical form. The abscissa is relative retardation (linearized inverse wavelength), from 0.5 to 1.5. The design range of the components is about 0.7 to 1.3 and is indicated by the vertical dashed lines. Note that the achromaticity of the m_{42} element is best for the Pancharatnam $Q \rightarrow V$ converter (solid curve), but that performance as a $V \rightarrow Q$ converter (m_{24}) is sacrificed for this; this converter is definitely *not* a modified quarterwave retarder, but it *is* a very good $Q \rightarrow V$ polarization converter.

Table 4.2 *Mueller matrix pictograms, illustrating the behaviour of ideal and multi-layer achromatized components*

Linear retarder

ideal (principal orientation) · (rotation matrix) · ideal (any orientation)

$$
\begin{pmatrix} 1 & 0 & 0 & 0 \\ 0 & 1 & 0 & 0 \\ 0 & 0 & R & R \\ 0 & 0 & R & R \end{pmatrix}
\quad
\begin{pmatrix} 1 & 0 & 0 & 0 \\ 0 & A & A & 0 \\ 0 & A & A & 0 \\ 0 & 0 & 0 & 1 \end{pmatrix}
\quad
\begin{pmatrix} 1 & 0 & 0 & 0 \\ 0 & AR & AR & AR \\ 0 & AR & AR & AR \\ 0 & AR & AR & R \end{pmatrix}
$$

Halfwave linear retarders

exact halfwave · Pancharatnam achromatic

$$
\begin{pmatrix} 1 & 0 & 0 & 0 \\ 0 & A & A & 0 \\ 0 & A & A & 0 \\ 0 & 0 & 0 & -1 \end{pmatrix}
\quad
\begin{pmatrix} 1 & 0 & 0 & 0 \\ 0 & A & A & (0) \\ 0 & A & A & (0) \\ 0 & (0) & (0) & (-1) \end{pmatrix}
$$

exact with dichroism · Pancharatnam with dichroism

$$
\begin{pmatrix} 1 & (0) & (0) & 0 \\ (0) & A & A & 0 \\ (0) & A & A & 0 \\ 0 & 0 & 0 & 1 \end{pmatrix}
\quad
\begin{pmatrix} 1 & (0) & (0) & (0) \\ (0) & A & A & (0) \\ (0) & A & A & (0) \\ (0) & (0) & (0) & (-1) \end{pmatrix}
$$

$Q \rightarrow V$ converters

exact quarterwave ($\eta = 45^\circ$) · two-material achromatic

$$
\begin{pmatrix} 1 & 0 & 0 & 0 \\ 0 & 0 & 0 & -1 \\ 0 & 0 & 1 & 0 \\ 0 & 1 & 0 & 0 \end{pmatrix}
\quad
\begin{pmatrix} 1 & 0 & 0 & 0 \\ 0 & (0) & 0 & (-1) \\ 0 & 0 & 1 & 0 \\ 0 & (1) & 0 & (0) \end{pmatrix}
$$

Pancharatnam quarterwave achromatic · Pancharatnam converter achromatic

$$
\begin{pmatrix} 1 & 0 & 0 & 0 \\ 0 & (0) & (0) & (-1) \\ 0 & (0) & (1) & (0) \\ 0 & (1) & (0) & (0) \end{pmatrix}
\quad
\begin{pmatrix} 1 & 0 & 0 & 0 \\ 0 & (0) & R & R \\ 0 & (0) & R & R \\ 0 & (1) & (0) & (0) \end{pmatrix}
$$

'Depolarizers'

ideal · rotating near-quarterwave · rotating near-halfwave

$$
\begin{pmatrix} 1 & 0 & 0 & 0 \\ 0 & 0 & 0 & 0 \\ 0 & 0 & 0 & 0 \\ 0 & 0 & 0 & 0 \end{pmatrix}
\quad
\begin{pmatrix} 1 & 0 & 0 & 0 \\ 0 & (0.5) & 0 & 0 \\ 0 & 0 & (0.5) & 0 \\ 0 & 0 & 0 & (0) \end{pmatrix}
\quad
\begin{pmatrix} 1 & 0 & 0 & 0 \\ 0 & (0) & 0 & 0 \\ 0 & 0 & (0) & 0 \\ 0 & 0 & 0 & (-1) \end{pmatrix}
$$

Element values are indicated as follows: 0; 0.5; ± 1; $-1 \leq A \leq +1$, a function mainly of component azimuth (orientation); $-1 \leq R \leq +1$, a function mainly of retardation (i.e. wavelength and component thickness); $-1 \leq AR \leq +1$, a function of both azimuth and retardation. Brackets indicate approximate values. The matrices have been normalized to a transmittance of unity for unpolarized radiation. Zero dichroism has been assumed except where explicitly stated. For spectral performance of the $Q \rightarrow V$ converters, see fig. 4.1.

Note: Personally, I welcome this particular extension of the nomenclature, although the notation should draw attention to the fact that complex visibilities are intended (e.g. by referring to 'Stokes visibilities' rather than Stokes parameters and using a script font for $\mathscr{I}$, $\mathscr{Q}$, $\mathscr{U}$ and $\mathscr{V}$; cf. section 6.2.3). Extension of Jones and Mueller calculus to radio (correlation) interferometers is a developing field; at the time of writing, I recommend Hamaker *et al.* (1995) and Sault *et al.* (1995).

4.2 Jones matrices and when to use them

When the *phase* of a polarized signal relative to some other polarized signal is important, the Stokes parameters are of no use; they deliberately ignore phase (except in a relative sense within each signal, as needed to specify the state of polarization). However, there are situations (such as when combining the beams of an interferometer) when phase does matter. Under such circumstances, one has to use *Jones vectors* and *matrices*, first introduced by R.C. Jones (1941); these represent the electric fields (and their transformations) of two orthogonal polarization forms (usually linear), including absolute phase if desired. Excellent modern presentations are given in Kliger *et al.* (1990, pp. 61–75) and Collett (1993, pp. 187–218). Since Jones calculus is not used a great deal in astronomy (except to design instruments), only the basic ideas will be presented here. The important thing is to know when it is necessary to use Jones rather than Mueller calculus (i.e. when phase is important) and where to find information in such a case. Typical advanced applications in astronomical instrumentation may be found in Chipman (1989), Chipman and Chipman (1989), November (1989), Sánchez Almeida and Martínez Pillet (1992) and in Sánchez Almeida (1994).

Jones calculus cannot handle partial polarization (mixed states of polarization). There are situations in which phase is important but polarization is partial. In such a case, one must formally separate the radiation into two (generally unequal) mutually incoherent fully polarized components (pure states) of orthogonal polarizations (i.e. that of the partial polarization in the input signal and its opposite), treat each separately by Jones calculus and obtain the final result by incoherent formal recombination of the outputs.

The notation used for Jones calculus is the complex notation for sinusoidally varying quantities. This notation is explained clearly in Born and Wolf (1964, pp. 494–9) and in Hecht and Zajac (1974, pp. 17–19 and 199). The crux is that one rewrites a cosine function as 'the real part of a complex exponential'. The advantage of the complex exponential is that the effect of a phase ϕ can be expressed as multiplication by $e^{i\phi}$ and that one can gather amplitude and phase of a signal into a single *complex amplitude* E:

$$\psi(x,t) = Re\left[Ae^{i(kx-\omega t+\phi)}\right] = Re\left[Ae^{i\phi} \cdot e^{i(kx-\omega t)}\right] \stackrel{\text{def}}{=} Re\left[E \cdot e^{i(kx-\omega t)}\right]$$

Knowing that, to obtain a physically meaningful quantity, we should multiply
E by the complex exponential $e^{i(kx-\omega t)}$ and then take the real part, we can add,
subtract, integrate, phase shift or otherwise transform E linearly. Of course,
the original wave function ψ represents an electric field and hence represents
the amplitude rather than the 'intensity' of the electromagnetic wave, which is
generally not an observable and in any case is never the desired end-product
of astronomical observations. To obtain the observable 'intensity' or the flow
of radiant energy, we must eliminate the absolute phase by squaring the
amplitude A, e.g. by obtaining $E \cdot E^*$, where E^* is the complex conjugate of
E. In the quasi-monochromatic and polychromatic cases, A and ϕ are ('slow')
functions of time; such 'slow' variations must be averaged in computing Stokes
parameters from complex amplitudes (see section 4.3).

A slightly different situation arises in (quasi-)monochromatic radio systems,
in which, by splitting a signal into two components and applying a 90° phase
shift to one of them, we can obtain both the real and the imaginary parts of
the signal as observables and do away with the mental *real part of* qualifica-
tion (the same process could of course be carried out at optical wavelengths,
but in the photon-limited conditions of astronomy there is usually no advantage to this). Jones matrix algebra applies whether the imaginary part
is an observable or not. For astronomical purposes one usually converts to
Stokes parameters and Mueller matrices at the end of the calculation (with
the added complication in correlation-type interferometers that the observables
are spatial cross-correlation products; these are complex quantities, viz. the
spatial complex Fourier transforms of the real Stokes parameter sky distributions, but they are nevertheless often referred to as Stokes parameters; see
section 6.2.3).

Jones represented a fully polarized signal as the vector sum of two electric
fields at right angles, as we did in words in chapter 2. In terms of complex
amplitudes,

$$\mathbf{E} = E_x \cdot \mathbf{l} + E_y \cdot \mathbf{m}$$

where $\mathbf{l}$ and $\mathbf{m}$ are unit vectors in the x and y directions. In the Jones calculus,
E is written as a column matrix, with complex components:

$$\mathbf{E} = \begin{pmatrix} E_x \\ E_y \end{pmatrix}$$

With every state of 100% polarization one can associate such a column matrix,
or 'Jones vector'. Distinction is made between 'full Jones vectors' (which include
amplitude and phase of both components) and 'standard normalized Jones
vectors' (for which the modulus is equal to unity). The standard normalized

Jones vectors for horizontally (along x) and vertically polarized radiation are, respectively,

$$\begin{pmatrix} 1 \\ 0 \end{pmatrix} \text{ and } \begin{pmatrix} 0 \\ 1 \end{pmatrix}$$

For linear polarization at position angle χ, the standard normalized Jones vector is

$$\begin{pmatrix} \cos \chi \\ \sin \chi \end{pmatrix}$$

For general elliptical polarization, the standard normalized Jones vector is

$$\begin{pmatrix} \cos \chi \\ \sin \chi \cdot e^{i\Delta} \end{pmatrix}$$

where Δ is the phase difference between the x and y components (in the sense $\phi_y - \phi_x$). More symmetrically, whenever absolute phase is not important, this is rewritten:

$$\begin{pmatrix} \cos \chi \cdot e^{-i\Delta/2} \\ \sin \chi \cdot e^{i\Delta/2} \end{pmatrix}$$

This form is in agreement with the sign of $kx - \omega t$ above, within the 1942 convention of the Institute of Radio Engineers (IRE, now IEEE; see Simmons and Guttmann (1970, appendix III)).

Circular polarization is represented by ($\chi = 45°, \Delta = 90°$):

$$1/\sqrt{2} \begin{pmatrix} 1 \\ i \end{pmatrix} \text{ and } 1/\sqrt{2} \begin{pmatrix} 1 \\ -i \end{pmatrix} \quad \text{or} \quad 1/\sqrt{2} \begin{pmatrix} -i \\ 1 \end{pmatrix} \text{ and } 1/\sqrt{2} \begin{pmatrix} i \\ 1 \end{pmatrix}$$

The Jones vectors

$$\begin{pmatrix} m \\ n \end{pmatrix} \text{ and } \begin{pmatrix} -n^* \\ m^* \end{pmatrix}$$

represent mutually orthogonal polarization forms. Any Jones vector may be expressed as a linear combination of *any such pair of orthogonal* Jones vectors; the standard normalized Jones vectors form a complete orthonormal set. In practice, pairs of orthogonal linear polarization forms ('horizontal' and 'vertical') are almost always chosen as the base of the complex vector space.

This completes the thumbnail sketch of the Jones *vectors*. To represent the action of a medium on polarized radiation, matrices are again employed. These matrices now have 2×2 elements, which, however, are complex. A non-polarizing and non-retarding absorbing medium absorbs both components equally and causes no relative phase shifts; it has the matrix

$$\begin{pmatrix} t & 0 \\ 0 & t \end{pmatrix}$$

Here t is the *amplitude* transmittance ($0 < t \leq 1$), while what we usually measure in astronomy is t^2, the intensity transmittance. For a linear polarizer with maximum transmittance for electric field vibrating along the x-axis, the matrix is:

$$\begin{pmatrix} t_x & 0 \\ 0 & t_y \end{pmatrix}$$

where $(t_x)^2$ and $(t_y)^2$ are the intensity transmittances for an electric field vibrating along the x- and y-axes, respectively ($t_x > t_y$). The term *diattenuation* is sometimes used to describe the phenomenon of $t_x \neq t_y$.

Just as with Mueller calculus, there is a rotator matrix:

$$\mathbf{T}(\theta) = \begin{pmatrix} \cos\theta & \sin\theta \\ -\sin\theta & \cos\theta \end{pmatrix}$$

A linear retarder (component with linear birefringence) produces *linear retardance*, a phase difference between the x and y components, but causes no relative absorption. The phase difference is $\Delta = 2\pi d(n_y - n_x)/\lambda$, the n-values being refractive indices and d being the thickness of the component. In principal orientation, i.e. with the fast axis of the retarder along the x direction, the Jones matrix of such a retarder is:

$$\mathbf{J}_{\Delta,0} = \begin{pmatrix} e^{i\Delta/2} & 0 \\ 0 & e^{-i\Delta/2} \end{pmatrix}$$

At angle θ, it becomes:

$$\mathbf{J}_{\Delta,\theta} = \mathbf{T}(-\theta) \cdot \mathbf{J}_{\Delta,0} \cdot \mathbf{T}(\theta)$$

For an extensive list of standard Jones matrices, see Kliger *et al.* (1990, appendix B).

Just as with the Mueller matrices, the order of the matrices is important (the matrices do not in general commute). Thus, one cannot easily establish the matrices for media which both retard and absorb within the same volume of space. For these applications, Jones developed an extension of his matrix calculus, essentially splitting the medium into an infinite number of infinitely thin layers; for an infinitely thin layer, one may think of the two actions taking place one after the other, and the order is immaterial (the infinitesimal matrices do commute; cf. the 'weakly polarizing optical train' in Stenflo (1994, section

13.4)). The infinitesimal result is then integrated to obtain the result for the entire medium. The differential matrices employed are known in the optical literature as Jones N-matrices; see Kliger *et al.* (1990, pp. 133–50).

Another specialized development of the Jones calculus, suitable for polarization ray-tracing of complete optical systems, is described in Chipman (1992, 1995) and in McClain *et al.* (1993); it uses three-dimensional complex electric field vectors, to allow for rays inclined to the system optical axis. Optical components are represented by 3×3 matrices with complex elements; the advantage for computer ray-tracing is that one system of coordinates can be used for the entire optical system, rather than many local systems and the transformations between them. For analysis and optimization of systems including polarization optics (e.g. Semel 1987), such polarization ray-tracing is indispensable, and it is being incorporated into optical design software (Chipman 1995).

4.3 Alternative definitions for the Stokes parameters

The definition of the Stokes parameters in section 2.2 was, for didactic reasons, given in terms of the axial ratio and azimuth of the polarization ellipse. These are not the quantities one actually measures, but the definition can be proved to be equivalent to forms couched in terms of measured quantities. The proofs are somewhat involved, and are most easily traced using van de Hulst (1957, p. 41), then Kliger *et al.* (1990, pp. 103–18, culminating in pp. 117–18); they should be read thoroughly once. The equivalent forms of the Stokes parameters are:

$$
\begin{aligned}
I &= I_0 + I_{90} = \overline{E_x E_x^* + E_y E_y^*} = \overline{A_x^2 + A_y^2} = \overline{a^2} \\
Q &= I_0 - I_{90} = \overline{E_x E_x^* - E_y E_y^*} = \overline{A_x^2 - A_y^2} = \overline{a^2 \cos 2\beta \cos 2\chi} \\
U &= I_{45} - I_{-45} = \overline{E_x E_y^* + E_y E_x^*} = \overline{2 A_x A_y \cos \Delta} = \overline{a^2 \cos 2\beta \sin 2\chi} \\
V &= I_{\mathrm{rc}} - I_{\mathrm{lc}} = \overline{i(E_x E_y^* - E_y E_x^*)} = \overline{2 A_x A_y \sin \Delta} = \overline{a^2 \sin 2\beta}
\end{aligned}
$$

where overlining denotes time- or ensemble-averaging. The first form is used in 'optical' polarimetry (from infrared to γ-ray), where equipment measures intensities and there is no simple way of obtaining cross-products of amplitudes; the second and third forms are more common in radio-polarimetry, where amplitudes and phases are available as physical quantities and (complex) amplitudes can be multiplied together in a 'correlator'. Actual measurement techniques are discussed in chapter 6.

The Wolf 'coherency matrix' formulation is equivalent to that using the Stokes parameters; it includes partial polarization. Sánchez Almeida (1992) discusses the equivalence in the context of radiative transfer, but otherwise it has not been used much in astronomy and will not be discussed here. Details may be found in Born and Wolf (1964, pp. 544–55), a short summary may be found in Stenflo (1994, section 2.6.2). The coherency matrix may be rearranged into a four-vector; its transformations and conversions are then represented by 4×4 matrices with generally complex elements. This formulation is related to the Stokes/Mueller formalism and is particularly suited to the treatment of radio (correlation-type) interferometers; the reader is referred to Hamaker *et al.* (1995) for details.

4.4 Complementarity of the Mueller and Jones representations

The Mueller and Jones formalisms (and related systems) might seem to be competing alternatives, but in fact they are complementary to a large degree. This section concerns the relationship between them and their respective niches in astronomical practice.

The addition of Stokes vectors represents *incoherent* combination of beams, or 'intensity superposition' (e.g. integrating the light from the visible hemisphere of a star); the addition of Jones vectors represents *coherent* combination of beams, or 'amplitude superposition' (e.g. the output of an optical interferometer or the signal at the focus of a radio telescope).

A number of relations connect the parameters of the Jones and Mueller vectors for (the polarized part of) a beam of radiation. They are listed here; the proofs may be found, for instance, in Kliger *et al.* (1990, table 5.2 and preceding pages). With $\Delta = \phi_y - \phi_x$ and $\tan \gamma = A_y / A_x$, the relations are:

$$\tan 2\chi = \tan 2\gamma \cdot \cos \Delta = U/Q$$
$$\sin 2\beta = \sin 2\gamma \cdot \sin \Delta = V/\sqrt{Q^2 + U^2 + V^2}$$

$$\tan \Delta = \tan 2\beta \, / \, \sin 2\chi = V/U$$
$$\cos 2\gamma = \cos 2\beta \cdot \cos 2\chi = Q/\sqrt{Q^2 + U^2 + V^2}$$

Systems that convert a 100% polarized input signal into a 100% polarized output signal can be described by Jones matrices. Their Mueller matrices can be derived from the Jones matrices; this derivation is sketched, and the result given, by van de Hulst (1957, p. 44; note particularly that the resultant Mueller matrices contain only seven independent constants). The constraints on such Mueller matrices are discussed clearly and succinctly by Hovenier (1994) (see also fig. 4.2); constraints such as these may become very useful in astronomy. Since such optical systems convert one pure state of polarization into another pure one, Hovenier refers to the Mueller matrices representing these systems

(a)

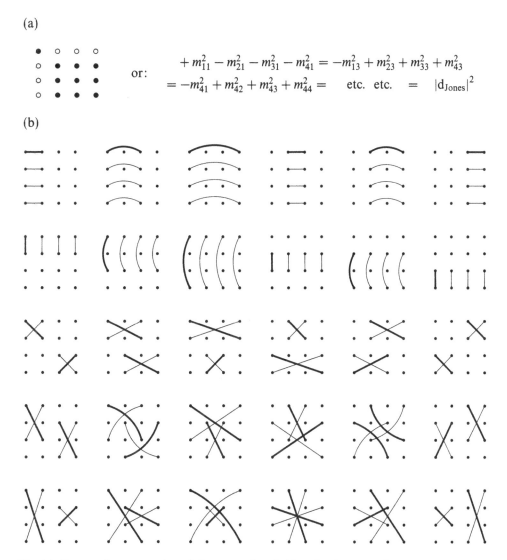

$$+ m_{11}^2 - m_{21}^2 - m_{31}^2 - m_{41}^2 = -m_{13}^2 + m_{23}^2 + m_{33}^2 + m_{43}^2$$

or:

$$= -m_{41}^2 + m_{42}^2 + m_{43}^2 + m_{44}^2 = \quad \text{etc. etc.} \quad = \quad |d_{\text{Jones}}|^2$$

(b)

Fig. 4.2 Systematic presentation of the constraints linking the elements of a 'pure' Mueller matrix; from Hovenier (1994) and references therein. (a) Seven relations between the squares of elements; dots in the pictogram represent positive squares of the elements of the Mueller matrix and circles represent negative squares; the sum of each row and also of each column of the pictogram is equal to the squared modulus of the determinant of the Jones matrix (or, equivalently, the positive square root of the determinant of the Mueller matrix). (b) Thirty relations between products of elements; each pictogram denotes one relation; the lines linking elements of the pictogram represent products of the elements of the Mueller matrix, a thick line for a positive product, a thin line for a negative one. The sum of the products within each pictogram is equal to zero, e.g. the top left pictogram represents $m_{11}m_{12} - m_{21}m_{22} - m_{31}m_{32} - m_{41}m_{42} = 0$, the bottom right pictogram $m_{21}m_{32} - m_{22}m_{31} + m_{13}m_{44} - m_{14}m_{43} = 0$.

as *pure Mueller matrices*; other terms in use are 'non-depolarizing', 'totally polarizing' and 'deterministic', none of which is as clear and concise as 'pure'.

Macroscopic astrophysical Mueller matrices may be thought of as ensemble averages or sums of (products of) microscopic matrices which themselves are 'pure' Mueller matrices to which the constraints do apply. Products of pure Mueller matrices are themselves pure (Hovenier 1994), but for sums of Mueller matrices this is not necessarily true (witness, for instance, the 'depolarizer' matrices of example 5 of section 4.1, which do not obey the 'sums of squares' rule of fig. 4.2(a)).

> **Note:** A system which has an associated Jones matrix can reduce the degree of polarization of *partially* polarized radiation, which is why 'non-depolarizing' is not a happy choice of terminology. A single example suffices to illustrate this (I am indebted to J.W. Hovenier for pointing this out to me): a *partial* polarizer (e.g. linear, Shurcliff (1962, p. 168)), presented with a *partially* polarized input signal of polarization orthogonal to its output for unpolarized light, will yield an unpolarized output signal; it is left to the reader to write this out as a Mueller matrix equation and to verify that for 100% polarized input the output is also 100% polarized.

Suitability of the Mueller and Jones calculi and of the 'three-dimensional Jones calculus' for polarization ray-tracing within telescopes and other instruments (optical, X-ray, radio) is discussed by Chipman (1992). Sánchez Almeida and Martínez Pillet (1992) use both Jones and Mueller calculi in their discussion of the polarization properties of optical telescopes. They use Jones calculus for deriving the effect of the optical system on fully polarized radiation, after which they convert to observables by deriving the Mueller matrix of the system as a function of image coordinates, pixel size and seeing conditions. In astronomy, which generally deals with radiation for which phase is irrelevant, this would appear to be the best procedure (perhaps even for some astrophysical calculations). In a similar vein, Hamaker *et al.* (1995) use Jones calculus to represent the response of one dipole of a single radio telescope with its associated receiver, then derive the Mueller matrix of a two-element correlation-type interferometer via the direct matrix product of the two Jones matrices for the individual telescopes; this promises to be a significant new departure in the description of the polarization aspects of radio (correlation-type) interferometers.

4.5 Use of the Poincaré sphere

Since the Poincaré sphere is used for a graphical representation of the Stokes parameters, it must be possible to represent a Mueller matrix graphically as some operation on or within the Poincaré sphere. Such a graphical method can

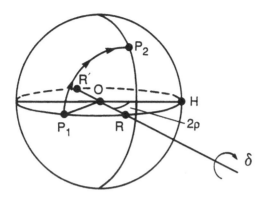

Fig. 4.3 Action of a retarder represented on the Poincaré sphere, adapted from Kliger *et al.* (1990).

be useful in the initial stages of analysing a problem; when the approximate solution has been found, it may be verified and perhaps improved by matrix calculus. The Poincaré sphere methods are described clearly in Shurcliff (1962) and Kliger *et al.* (1990); the latter also gives most of the proofs. For more detailed proofs and further illustrations, consult Ramachandran and Ramaseshan (1952) and Jerrard (1954). Examples of the design of polarization components by using the Poincaré sphere may be found in Pancharatnam (1955a,b) and Koester 1959; the results of matrix calculations on Pancharatnam's original designs are included in table 4.2 and fig. 4.1. Landi Degl'Innocenti and Landi Degl'Innocenti (1981) use the Poincaré sphere in an astrophysical application, constructing an analogy between polarized radiation transfer and the motion of a charged particle in the presence of electric and magnetic fields. The essence of these graphical methods is as follows:

- Any point R on the Poincaré sphere represents a particular state of polarization.
- The opposite end of the diameter through R represents the orthogonal polarization R'.
- Any homogeneous birefringent medium (a 'retarder') has two eigenstates of polarization, those states of polarization for which the radiation can propagate through the medium without change of polarization. These eigenstates are orthogonal.
- The diameter RR' therefore represents the medium with eigenstates R and R' (hence the choice of R, for 'retarder'). R is taken to be the fast eigenmode, i.e. that with the lowest refractive index.
- Given an initial polarization state represented by the point P_1 on the Poincaré sphere, the action of the retarder (retardation δ) is to produce an output

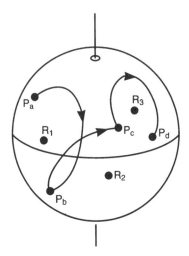

Fig. 4.4 Action of several (elliptical) retarders in series; from Shurcliff (1962), reprinted by permission of Harvard University Press.

state of polarization P_2, where P_2 is obtained from P_1 by striking an arc of length δ from P_1 along a (generally small) circle *centred* on RR' (or rotating the sphere and its coordinates around RR', whilst keeping the point P_1 fixed in space). The arc is struck *clockwise*, or the equivalent sphere rotation is anticlockwise, when looking back from infinity at the R end of the diameter (fig. 4.3); if R had been chosen to be the slow eigenmode, the arc and the rotation would have been reversed.

- Several retarders in series are represented by several rotations in succession, of appropriate size and about appropriate axes. The general case of elliptical retarders acting on elliptical input polarization is illustrated in fig. 4.4.

- If light of polarization state P passes through a polarizer of state A (A for 'analyser'), then the intensity of the light transmitted is $\cos^2(\widehat{PA}/2)$, where $\widehat{PA}$ is the length of the *great* circle arc between P and A. The proof of this is outlined in Ramachandran and Ramaseshan (1954, pp. 51, 52) and two examples are of particular interest:

  – When $\widehat{PA} = 90°$, the transmission is 0.5; thus any linear polarizer transmits half of any incident circularly polarized radiation, and a circular polarizer transmits half of any incident linearly polarized radiation (not counting any polarization-independent attenuation).

  – When $\widehat{PA} = 180°$ (orthogonal forms), the transmission is zero; for every (in general elliptical) 100% polarizer, a corresponding (orthogonal elliptical) analyser exists which yields zero transmission for the combination of the two.

4.6 The complex plane of polarization states

Like the alternative matrix calculus in terms of complex amplitudes of the electric fields, there exists an alternative graphical representation in terms of a complex plane; like the Jones calculus, it cannot be used for partial polarization. The complex quantity represented in that plane is $E_y/E_x = (A_y/A_x)e^{i(\phi_y-\phi_x)}$. The relationship between this complex plane and the *surface* of the Poincaré sphere is by stereographic projection, which has the useful property that circles project into circles; the relationship is explained and illustrated in Kliger *et al.* (1990, pp. 118–33) and in Collett (1993, pp. 237–44). Like the Jones calculus, the complex-plane representation is unlikely to be needed by the practising astronomer (as opposed to the instrument designer).

4.7 Astrophysical use of Mueller matrices

For almost all conceivable astronomical situations, the Mueller matrix is the most general description of the processing of polarized radiation. Whenever the polarization is expected to provide useful astronomical information, we should describe the propagation of the radiation from source to detector by Mueller matrices. If we do not know all the elements of these matrices, the proper course is not to hide our ignorance by collapsing the matrix to its top left element, but rather to examine how all the elements enter into the desired final result and only then to decide whether to neglect them or to make an effort to determine them by measurement or theory. We should of course make use of whatever knowledge we have of the symmetry properties or other constraints of the matrices involved.

Some of the Mueller matrices used in applications are similar to Jones N-matrices (section 4.2). They are 'differential matrices', i.e. they express the modification, *per unit path length*, of the Stokes parameters due to the properties of the medium and they are functions of position within the medium. Discussions of polarization radiative transfer (for instance with the aim of deducing stellar magnetic fields from the complicated Zeeman profiles one observes in the integrated light from a star) may be found in Landi Degl'Innocenti and Landi Degl'Innocenti (1981), in Landi Degl'Innocenti (1992) and in Stenflo (1994). The reader new to polarization concepts is referred to Rees (1987) for a 'gentle' introduction, followed by Landi Degl'Innocenti (1987); the latter enumerates (pp. 266, 267) the physical meaning of the elements of the Mueller matrix; after such preparation, the reader may graduate to Casini and Landi Degl'Innocenti (1993) and to Stenflo (1994). Deguchi and Watson (1985) use polarized radiative transfer in the interstellar medium to interpret circular spectro-polarimetry in the 21-cm H and 18-cm OH lines, in terms of the Zeeman effect and possible linear-to-circular polarization conversion. For the case that depolarization of 100% polarized radiation does not occur (as is specifically mentioned by Landi Degl'Innocenti and Landi Degl'Innocenti (1981) and is implicit in Sánchez Almeida (1992)), the differential matrix has only seven independent parameters and the cumulative matrix for an entire optical path is then a pure Mueller matrix (i.e. it has an associated Jones matrix; see Sánchez Almeida (1992) and references therein; Stenflo (1994, section 2.6.2) is helpful when reading this paper). When depolarizing processes are important (e.g. scattering of radiation from other volume elements of the medium), the Mueller matrices will, in general, not be pure; they must, however, satisfy the 'Stokes criterion', i.e. for an arbitrary input Stokes vector they must yield a physically possible output Stokes vector (with a degree of polarization not exceeding unity).

Another area of application with a considerable history is the scattering of sunlight within planetary atmospheres (a related case is scattering of photospheric light in a circumstellar shell, e.g. Voshchinnikov and Karjukin (1994)). Instead of reflection, transmission and scattering *coefficients*, one should use the corresponding Mueller *matrices* (see, for instance, Stammes *et al.* (1989)). Multiple scattering is the rule in such situations, and rotation matrices are needed to transform from each local scattering coordinate system to the next; large errors are likely if one uses scalar rather than matrix calculations in multiple scattering (see section 5.7). When the atmosphere is of noticeably less than infinite thickness, the polarization properties of the planet's solid surface enter into the problem; a measurement, guess or estimate of many of the elements of the reflection matrix of the surface is then required, while in fact usually no more than four (or even one) of these have been determined. The reader is referred to Van de Hulst (1980, chapter 15 ff.).

A third example is distributed generation of synchrotron emission and subsequent Faraday rotation, these two mechanisms overlapping in space. The traditional treatment has been very simple: the transverse component of the magnetic field is considered for the synchrotron emission, while only the longitudinal component is considered relevant for the Faraday rotation. In fact, these are convenient approximations (isotropic velocity distribution of relativistic electrons; circular birefringence only) which are not always valid, in particular when relativistic mass motions destroy the symmetry so conveniently assumed. Mueller matrix treatment therefore becomes necessary; it is complicated (see Jones and O'Dell 1977 and Jones 1988), but computer-aided intelligence is bound to provide insights, at the very least about the uniqueness of models 'derived' from the observations.

5

Instruments: principles

In this chapter, instrumental principles will be discussed, with emphasis on system behaviour and without any preconceptions about the wavelength at which one observes. Practical illustrations will inevitably relate to a particular wavelength region (optical and radio, which is where the experience resides). It may therefore be necessary to scan chapter 6 before attempting to understand the present chapter in detail.

5.1 Telescopes

The first optical element of an astronomical observing system is always a telescope (disregarding the atmosphere for the present discussion). It is important to realize that, in general, a telescope will modify the polarization of the radiation before the polarimeter measures it. It is equally important to have some general feeling for the conditions under which such modification is likely to be appreciable and how it can be minimized.

The guiding principle is symmetry; any departures from full symmetry will modify the polarization. The considerations below illustrate this, but full understanding will require mathematical treatment by Mueller or Jones calculus, with optical constants applicable to the wavelength of interest.

Oblique incidence on a mirror produces both diattenuation (polarizing action) and retardation (wave plate action). These effects are minimal at near-normal and, somewhat surprisingly, at grazing incidence; the largest effects occur at intermediate angles of incidence, the details depending on the values of the real and imaginary parts of the refractive index (which in their turn depend on the wavelength). In general, Coudé and Nasmyth telescopes will exhibit strong polarization modification, while prime-focus, Cassegrain and Gregorian systems will be relatively free from it.

Note: This statement also applies to grazing-incidence X-ray telescopes. There has been much confusion in the literature about phase changes at grazing incidence (at any wavelength), both at a conducting surface and for total internal reflection within a dielectric. This confusion is due to (mis)interpretation of the sign of the amplitude reflection coefficients, in relation to the coordinate systems for the electric field vectors of the incoming and reflected rays. The relationship between these two coordinate systems for normal incidence reverses for grazing incidence, which means that a formal retardance value of 180° at grazing incidence must not be interpreted as halfwave action, but rather as zero retardance. Experiment has indeed shown that incident circularly polarized light does not change its handedness on reflection at grazing incidence.

Rotationally symmetric telescopes of large focal ratio (slow optical systems) show very little *linear* polarization of unpolarized incident radiation for an image on the optical axis, since the polarizing action of different parts of the mirror(s) is basically radially oriented and the resultant averages out.

Note that even rotationally symmetric telescopes are *not* ideal, they do not transmit the state of polarization of incident radiation to the on-axis focal image without any change whatsoever: a certain amount of net (pseudo-)de-polarization of the incident radiation will remain after the averaging; the converted linear polarization averages out to zero, with the result that the diagonal elements are less than unity. This is not all, however: contrary to the lowest level of intuition, the off-diagonal elements do not all average to zero, either; what remains is (rotationally invariant) coupling between Q and U and separately between I and V (but *not between these pairs* of Stokes parameters; see McGuire and Chipman (1988), particularly J.O. Stenflo's foreword to that report, which makes clear that with a more sensitive intuition we should really have expected a result of this nature).

Note: When the telescope aperture is sufficiently small in terms of the wavelength of the radiation, the diffraction pattern will extend beyond the image determined by geometrical optics, and a single point in the focal plane (e.g. the on-axis image) will receive radiation from more than one direction in the sky: the spatial point spread function (antenna pattern) will have sidelobes, which have polarization properties of their own. This is a problem mainly in the radio region of the spectrum and is discussed in section 5.5.5.

For off-axis images the rotational symmetry is broken; the angles of incidence on different parts of the mirror(s) are no longer symmetrically distributed, so that residual polarizing action will exist, larger for faster (lower focal ratio) optical systems. Schmidt *et al.* (1993) discuss a 'software beam-switching' system for a wavelength of 2.8 cm, and in their very instructive figure 4 show (linear) polarization antenna patterns for each of the four offset focus antennas. The focal-plane antennas of the VLA and VLBA radio telescopes are offset, in order to allow front-ends for several frequencies to be mounted permanently. This introduces asymmetry, instrumental polarization and a different primary antenna pattern for the two orthogonal polarization channels of the receivers (the latter feature causes problems in obtaining the highest polarization accuracy from these synthesis arrays; the Westerbork telescope does not suffer from this, but requires more complicated operations to change frequency); the matter is discussed and references are given in Spoelstra (1992). An optical case is discussed by Sánchez Almeida and Martínez Pillet (1992), who conclude that only fast, wide-field systems are likely to show measurable effects in the optical region (cf. Schmidt *et al.* (1992), section 2.1, on the original Multi-Mirror-Telescope).

The design of 'polarization-free' telescopes is becoming an important part of astronomical engineering. Any oblique reflection that is not symmetrical about the optical axis should be avoided or made innocuous by preceding it with a polarization modulator (see section 5.2; the LEST design of fig. 5.1 is an example of the latter approach, as is the fibre link from Cassegrain focus to Coudé spectrograph mentioned by Donati *et al.* (1992)). Alternatively, but less fundamentally, some kind of compensation may be used (e.g. Martínez Pillet and Sánchez Almeida 1991). The ideal solution is a full-aperture modulator as the first optical element of the telescope; the only such system that I know of is WISP, a far-ultraviolet Schmidt telescope of 20-cm aperture with a rotatable wave plate as its entrance window (Nordsieck *et al.* 1994b).

For short exposures, atmospheric seeing also breaks the rotational symmetry of the system: bright patches in the system pupil cause different weights for different parts of the mirror surface, so that the polarizing properties of the different sections of the telescope mirrors no longer cancel. Sánchez Almeida (1994) investigates this in detail and concludes that some of the off-diagonal elements ('instrumental polarization' and 'polarization conversion', see section 5.5) of the normalized Mueller matrix of *atmosphere plus telescope (upstream of modulator)* may reach 'instantaneous' values of ($\pm$) a few tenths of a per cent (all-reflecting telescope) to a few per cent (realistic entrance window for an enclosed telescope) at individual speckles within the seeing disc. When averaged over the seeing pattern or over long periods, the polarization effects disappear. These findings will be of practical importance only for high-resolution polarimetry of bright sources (sufficiently low photon noise for the optical polarization errors to dominate, even in speckles). Solar speckle polarimetry (Keller and von der Lühe 1992) is an immediate concern, and the infrared may well be the wavelength region of prime importance. A conceptual cure for such effects would be a full-aperture rapid polarization modulator, since that would code only the polarization of the incoming signal and render

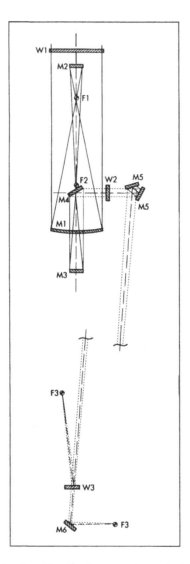

Fig. 5.1 A virtually polarization-free Coudé telescope; the optical design for the Large Earth-based Solar Telescope LEST (from Engvold 1992). The optical system for this helium-filled telescope includes windows (W) and mirrors (M); focal positions (F) are also indicated. The polarization modulator is mounted at the focal position F2, while the first oblique reflection is at M4.

the polarization properties of the telescope irrelevant; future developments in liquid crystal technology will bear watching in this respect, particularly for a telescope with an integral entrance window (fig. 5.1).

Even when the telescope is rotationally symmetric and we are observing the complete seeing disc as a single entity, on-axis, we may expect some polarizing action by the mirror surfaces, owing to technical imperfections constituing a lack of symmetry of the internal structure of the surface. Early aluminium coatings for optical mirrors sometimes showed strong polarization, which could also depend strongly on wavelength. This was traced to asymmetric and too rigorous cleaning of the surface before coating and/or to oblique incidence of the aluminium atoms during the coating process; it is generally not a significant problem in modern telescopes. Similar effects are to be expected from internal details of the mechanical support structure or the reflecting surface of radio mirrors; at short X-ray wavelengths, structure in the 'mirror' at the atomic-lattice level may be expected to cause trouble, while at longer wavelengths harmful residual structure could remain after diamond-machining to produce the complicated shapes required for X-ray mirrors. The perennial questions will be *to what extent* a mirror is ideal and what else it does.

The analysis of polarimetric errors usually deals with the entire optical system, i.e. including the telescope. Indeed, in synthesis telescopes it would be difficult to say where the 'telescope' ends and the 'instrument' begins.

5.2 Modulation

In many astronomical applications, the polarized radiation flux is only a small part of the total. Under such circumstances, small errors in flux measurement could lead to large fractional errors in the degree of polarization. Often this can be calibrated, but such calibration uses valuable observing time, and the errors may vary too rapidly for calibration to be feasible. The problems include flexure of instruments or their supports, magnetic fields influencing detectors, dewing of optical surfaces and of feed antennas, etc.; they will not be discussed here, since a near-perfect cure exists in the form of the modulation technique.

The technique of modulation makes the measurement of *degree of polarization* (or the *normalized Stokes parameters* $Q/I, U/I, V/I$) insensitive to many errors. Basically, modulation is a way of making a differential measurement very rapidly and is useful whenever the required information is represented by a small quantity superposed on a large, irrelevant background signal. For polarimetry, modulation is implemented as a rapid switching of the polarimeter sensitivity between two orthogonal states of polarization and measuring the ratio of the alternating ('AC') signal to the average ('DC') signal. This ratio is proportional to Q/I, U/I, V/I or some combination of these, depending on the adjustment of the instrument. Clearly, such ratios are insensitive to external effects which multiply AC and DC by the same factor, such as time-varying gain of a radio receiver or scintillation in the atmosphere. Any (drift in the) zeropoint error will also be reduced; it will not affect the AC component, and the fractional error in degree of polarization will only be that of the I measurement.

A basic requirement for modulation techniques to work is that the modulation is faster than the time-constant of the errors which one tries to eliminate. For scintillation noise to be eliminated from optical polarimetry, modulation frequencies should be tens or hundreds of hertz; for elimination of slow gain changes (flexure of spectrographs, changing humidity in a radio feed, etc.), modulation may only be at millihertz frequencies (in this case, one could equally well call it differential measurement; the principle is the same).

5.3 Correlation

Radio techniques make use of polarized antennas in the focal plane of the telescope. Such a polarized feed antenna converts the free-space field of *one particular polarization form* into an electrical signal within a receiver system. It is possible to mount a pair of feed antennas of orthogonal polarizations within one focal-plane structure and to allow these to feed two entirely separate receivers (including the detectors, which square the amplitude signal to produce an output proportional to incident radiant energy). Depending on the polarization and/or orientation of the feed antennnas, the difference of the 'energy' signals in these two channels represents Q, U or V, while the sum represents I. However, the two receivers, though as far as possible identical, in fact differ by small amounts which may vary with time, so that the looked-for small differences between large signals may be unstable. For that reason, the 'sum and difference' method is not the most widely used in the radio part of the spectrum. Instead, one makes use of the fact that a polarized wave generates signals with correlated (complex) amplitudes in a pair of orthogonal polarization channels (cf. section 4.2; see also section 6.2.2). Such systems are not immune from error, but the recorded polarized flux is proportional to receiver gain acting on the incident *polarized* flux rather than to small differences in gain acting on the *total* flux. The recorded signal is therefore more stable to fractional gain changes by a factor of order $1/p$. However, stray signals that are correlated in the two channels will show up as spurious polarization; an example of this is mixer noise in one channel, introduced also into the other channel by non-orthogonality of the feed antennas or imperfections of the waveguide or horn in which they are mounted.

5.4 Statistics of polarization parameters

In reducing or modelling polarimetric observations, one does have to be careful of one's statistics. Fig. 5.2 shows that degree and angle of polarization will not have a Gaussian distribution when the degree of polarization is small, even

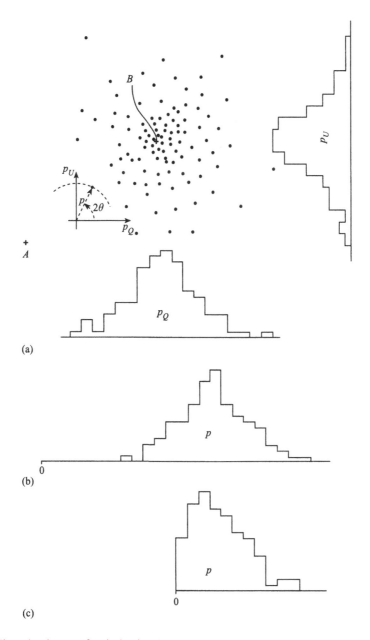

Fig. 5.2 When the degree of polarization is small compared with the measurement noise, the distribution of both degree and angle of polarization become noticeably non-Gaussian. (a) A set of measured p_Q and p_U, with histograms representing their distributions. The shape and width of these distributions are invariant with respect to choice of polarization (vector) zeropoint (only $\overline{p_Q}$ and $\overline{p_U}$ change). Parts (b) and (c) show histograms for p with different signal-to-noise ratios $\overline{p}/\sigma_p$. (b) $\overline{p_{(Q,U)}} \gg \sigma_{p_{(Q,U)}}$ (zeropoint at A); this distribution will approach that of p_Q and p_U as A moves away to infinity. (c) $\overline{p_{(Q,U)}} \ll \sigma_{p_{(Q,U)}}$ (zeropoint at B). The distribution of θ similarly changes character as the polarization zeropoint moves from B to A to infinity. For $\overline{p_{(Q,U)}} \gg \sigma_{p_{(Q,U)}}$, $\sigma_{p_Q} = \sigma_{p_U} \approx \sigma_p \approx 2p\sigma_\theta$ (θ in radians).

though Q and U (or Q/I and U/I) may have a distribution which departs very little from a Gaussian. This fact implies that one should not average degree and angle of polarization; instead, one should convert them to Q and U or Q/I and U/I, then perform the averaging, and finally convert back to degree and angle.

In some cases, a *normalized* Stokes parameter (Q/I etc.) is determined directly, as a 'noisy signal' within the polarimeter electronics or online software. This noisy signal is then averaged to reduce the noise. For a low signal-to-noise ratio of the individual measurements (below about 30 for the signal-to-noise ratio of Stokes I, which is always in the denominator), one should allow for the fact that the *normalized* Stokes parameters themselves do not have a Gaussian distribution. Details are given in Clarke and Stewart (1986), who discuss a number of statistical points peculiar to polarimetry (optical applications are assumed, but most of the paper is applicable to all spectral regions); Maronna *et al.* (1992) further refine the estimation of the normalized Stokes parameters.

The most likely occasion on which one might inadvertently break the above rules is when rebinning observations from detector pixels to some more meaningful coordinate such as wavelength or arcseconds on the sky; it should be remembered that rebinning is weighted averaging and must be performed as described above.

Subtle statistical questions involved in establishing reliable astronomical polarization standards are discussed by Clarke *et al.* (1993) and by Clarke and Naghizadeh-Khouei (1994).

5.5 Instrumental polarization errors and their calibration

Since astronomical degrees of polarization are often small, it is fortunate that many of the errors affecting photometry do not trouble polarimetry. Because the required quantities depend on ratios of fluxes at one (effective) wavelength, many errors tend to cancel out as long as these fluxes are measured at the same time by the same equipment. This is why accurate polarimetry of bright point sources can often be carried out during observing conditions that are too bad for almost any other type of observation.

However, there are several specifically *polarimetric* errors which must be discussed; each of them may be a function of wavelength and/or position within the image or pupil. Such errors include (vector) background polarized flux, scale error in degree of polarization, zeropoint error in polarization angle and degree-of-polarization (vector) zeropoint. At radio wavelengths, one should add the errors due to polarization sidelobes to this list. In this section I shall discuss generalities; for more detail, the reader should consult instrument manuals or reviews of instrumentation, e.g. Tinbergen and Rutten (1992, pp. 11–19) for optical CCD spectro-polarimetry, and Weiler (1973), Thompson *et al.* (1986) and Spoelstra (1992) for radio (synthesis) polarimetry. The hybrid case of (sub-)millimetre polarimetry is illustrated very well by Clemens *et al.* (1990).

In a photometric system, the equation relating the true source flux I entering the telescope to the recorded signal i might in a simple case look something like this:

$$i = i_0 + G_{instrument} \cdot \{I_{dome} + G_{telescope} \cdot (I + I_{background})\}$$

where the G factors represent 'gains', I_{dome} includes any sort of radiation that bypasses the telescope, $I_{background}$ includes man-made and natural sources of 'sky brightness' and i_0 is an electronic zeropoint error. By judiciously nodding the telescope or chopping with the secondary or a succeeding mirror, by sky and standard-source observations, and by assuming constancy of the equipment errors for a certain time or within a certain regime, one can calibrate many of these errors and, to a large extent, remove them. For polarimetry, of course, the above equation will become:

$$\mathbf{s} = \mathbf{s_0} + \mathbf{M_{instrument}} \cdot \{\mathbf{S_{dome}} + \mathbf{M_{telescope}} \cdot (\mathbf{S} + \mathbf{S_{background}})\}$$

which might look quite complicated when written out in full, since the **s** and **S** are Stokes vectors and the **M** are Mueller matrices. Fortunately many elements of the matrices vanish or can be parametrized, and in many cases either linear or circular polarization can be neglected. This leads to the simpler types of error discussed in the following, which generally suffice to describe the situations encountered in present practice. However, as treatment of polarization both in astrophysics and in instrumentation becomes more sophisticated, one must expect other and more subtle kinds of error to emerge. Matrix analysis will be the proper tool for describing these errors and for devising suitable calibration schemes (see, for example, McGuire and Chipman (1988), Elmore (1990), Xilouris (1991), McKinnon (1992a) and Sault *et al.* (1995)).

The present situation is that optical and single-dish or phased-array radio systems are often analysed in terms of Mueller and/or Jones matrices. The usual procedure is to derive first the Jones matrix for individual rays through the entire system; at the end of the calculation one converts to Mueller matrices for astronomical application, averaging over the optical pupil for each point in the focal plane. Synthesis instruments have so far not been treated by these matrix methods (but this will change, as outlined by Hamaker *et al.* (1995)). Aperture synthesis provides extra freedom in devising schemes to calibrate the system: the equipment of one individual telescope is largely independent of that of the other telescope in the elementary interferometer pair. Often the dipoles in one telescope can be oriented at any desirable angle with respect to those of the other one (mechanically or by electronic processing), and the connections between dipoles and electronics can be swapped between channels. In general, each synthesis instrument has its own folklore in what measures are actually taken to generate a simulated 'perfect' instrument from imperfect components. Matrix analysis will be a valuable tool in identifying the instrumental parameters that are important for polarimetry; using such insight, one may then set up optimal calibration schemes. See Sault *et al.* (1995) for such a discussion in matrix terms.

5.5.1 Polarized background

In astronomy one is often concerned with point sources which appear against a smooth background such as moonlit sky or galactic radiation. To eliminate this background, one subtracts a separate background observation from the 'source + background'; in polarimetry, one should do this for each of the four Stokes parameters, before obtaining the quotients $Q/I, U/I$ and V/I. When the polarized flux of the sky background is a function of position (e.g. scattered moonlight), it may be necessary to average background measurements at several positions. In radio synthesis polarimetry, the background (in Stokes *I and* the other Stokes parameters) can be eliminated by leaving out the short-spacing interferometers (often this is unavoidable, particularly in the case of the near-zero spacings: see section 7.5); the drawback is that this procedure amounts to spatial high-pass filtering, and low spatial frequencies in the true sky distribution are rejected at the same time.

5.5.2 Polarization angle reference

Operationally, polarization angle is measured with respect to some *instrumental* zeropoint. This is quite adequate as long as calibration observations relate the instrumental zeropoint to a reference direction on the celestial sphere. In some cases, relative polarization angles may be all that one needs; in such cases, calibration of the instrumental zeropoint may not be necessary (cf. most astronomical photometry, which in the absolute sense is less accurate than 1%, yet astrophysical information is obtained at the 0.1% level or even better).

Specialized methods have been developed to tackle calibration of polarization angle. At optical wavelengths, a reversible Polaroid can be suspended in front of a nearly horizontal telescope (Gehrels and Teska 1960), and one could devise methods with a water or mercury pool, while the scattering polarization from a low-albedo asteroid (D. Clarke, private communication, 1991) or from the terrestrial blue sky at the zenith (Können *et al.* 1993) have both been suggested. Absolute accuracy of a few tenths of a degree should be possible. At radio wavelengths, the usual practice is to align the dipoles within the feed antenna optomechanically and assume that the feed waveguide structure does not rotate the direction of vibration; orthogonality of the two dipoles within one feed is also tested electronically. The resultant accuracy is considered to be a few tenths of a degree. A specially constructed calibration transmitter on a satellite would be one way to improve accuracy still further, as would (at some wavelengths) a wire-grid polarizer installed in the telescope. At any one wavelength, one may adopt a celestial source as one's standard and relate all installations at this wavelength to each other. However, polarization angles at one wavelength relative to those at another will remain uncertain to some extent; this could be of importance in high-accuracy studies of Faraday rotation or within-source magnetic field structure.

5.5.3 *Degree-of-polarization scale or polarimetric efficiency*

An ideal polarimeter has a polarimetric efficiency of unity, i.e. a polarization of $x\%$ in the input is actually recorded as a polarization of $x\%$. Practical polarimeters have to make do with efficiencies less than unity in many cases. The value 'one minus the polarimetric efficiency' is often referred to as the 'instrumental depolarization'; insofar as different components added at the output have been combined with random phases (e.g. different wavelengths within the passband), this represents true depolarization, but to the extent that the process of superposition is coherent (rays within the beam of an optical polarimeter), 'pseudo-depolarization' would be a more accurate term (cf. section 4.1, example 5).

The usual cause of reduced polarimetric efficiency is the fact that retarders or analysers do not work perfectly at all wavelengths included within the instrumental passband. For this reason, it is mainly a problem in optical polarimeters; radio systems manipulate relative phases in local oscillator signals to simulate retarders with inherently broadband characteristics, while dipoles are sufficiently perfect polarizers over wide passbands.

The normal method of dealing with imperfect polarimetric efficiency is to calibrate it by observing some standard source (or any suitable source through some standard component) or by injecting a standard signal. Standard sources must be calibrated themselves. Optical standard components are polarizers; if they are not perfect, they can sometimes be made so in practice by cascading

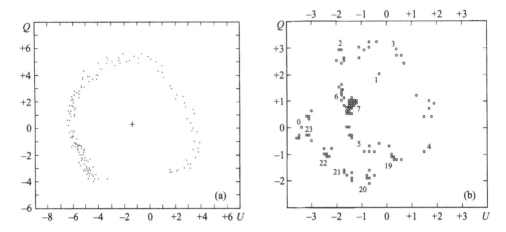

Fig. 5.3 Calibration observations over 12 hours (or more) of one point in the sky (at 1411 MHz and in the alt-azimuth frame), during (a) night-time, (b) daytime. The centre of the circle in (a) identifies the non-zero instrumental linear polarization zeropoint of the system. As is obvious when one traces the numbered chronological sequence of points in (b), the daytime 'circle' is badly distorted by (probably solar) radiation entering through the sidelobes. Note that the Q- and U- axes are interchanged compared to the usual practice. From Spoelstra (1972a,b).

them. Radio calibration of polarimetric efficiency is effected by injecting correlated noise into both channels at equivalent points of the receivers; this, of course, fails to calibrate the very first part of the chain, i.e. the telescope and feed structure, the dipoles and the first lengths of cable.

Since astronomical polarizations are usually much less than 100%, one is in danger during calibration of either overloading the hardware designed to handle the polarized signal or of losing accuracy by too small a signal in the 'total signal' channel. Sometimes there are technical solutions to this, such as the 'pile-of-plates' polarizer in the optical region, which can produce a calibrated 5 to 10% polarization in the light of a bright unpolarized star.

Polarimetric efficiency is often a function of wavelength or position in the field of view, and such functional dependence must be calibrated as well. The details will depend on the instrument and on the application.

5.5.4 Instrumental zeropoint of (degree of) polarization

Most telescopes and instruments polarize radiation to some small extent; this property is generally called *instrumental polarization*, but *zeropoint of (degree of) polarization* is a more descriptive term. When one observes a source of zero polarization, one generally obtains some significant output in Q, U and V. As long as this polarization is small, it is (vectorially) added to the true polarization signal when one observes any other source, and it may be (vectorially) subtracted during the reduction; whether the vector addition is to the Stokes vector itself or to the vector degree of polarization depends on the situation (e.g. point source or background, and cause of the zeropoint shift). If both the polarization of the signal and the zeropoint are large, full Mueller matrix treatment becomes necessary.

Such vector zeropoints are determined by observing sources of (near-)zero polarization, if these are available. When they are not (and one can rarely be sure a given source actually has zero polarization), one uses randomly selected ensembles of sources of low polarization (e.g. the average of the nearest 100 stars, excluding those with 'funny' spectral types, which might be intrinsically polarized; to get a flavour of such efforts, see the discussions in Clarke *et al.* (1993) and in the papers they cite).

The most fundamental method of eliminating the instrumental zeropoint of *linear* polarization is to use a telescope of the alt-azimuth type, using the fact that the sky rotates with respect to the telescope to determine and eliminate the instrumental linear polarization. Fig. 5.3(a) shows observations of a polarized celestial source during a period of 12 hours. In the instrumental (alt-azimuth) frame, a circle is described in the Q, U (or Q/I, U/I) plane; the centre of this circle represents the instrumental polarization to be removed. Clearly, if one can trust the constancy of this circle, the average of two observations at opposite ends of a diameter is a reliable estimate of the instrumental polarization. Optically this can often be used: one averages (in the instrumental frame) two observations of the source, timed so that between the observations the sky has rotated exactly 90° with respect to the telescope (it is best if the two observations are also symmetrically arranged with respect to the meridian, since any flexure of the telescope or instrument should then be identical in both and true source polarization can be obtained; that same source can then be used to investigate whether the system zeropoint varies with elevation). With radio observations, ionospheric Faraday rotation may invalidate this simple procedure: one needs enough points on the circle to define the centre without any information as to where on the circumference the point has been moved to by Faraday rotation. Fig. 5.3(b) shows that the assumption of constancy of the 12-hour circle is not always warranted; the cause may be solar radiation in the sidelobes, flexure of optomechanical assemblies or any other change of ambient conditions. McKinnon (1992a) documents polarization calibration of a phased array, using the alt-azimuth mount of the individual telescopes to good effect.

In synthesis radio telescopes, the instrumental zeropoint is a function of position within the field, due to the polarization sidelobes of the individual dishes (Napier 1989 and Wieringa *et al.* 1993). It can be strong, but in principle

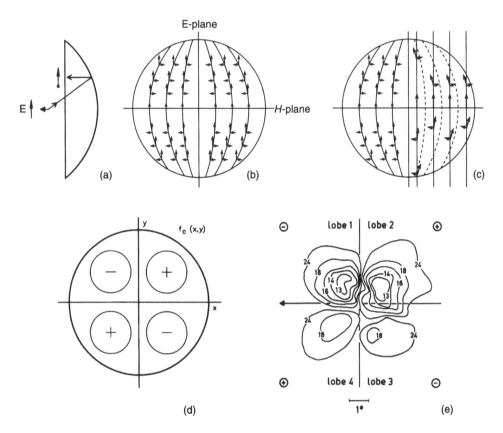

Fig. 5.4 Cross-polarization sidelobes, adapted from Westerhout *et al.* (1962) and Napier (1989, pp. 56–7); showing the cause, the schematic field pattern and a real-life case of such sidelobes. In the top half of this figure, (a) and (b) show the situation for transmission from a focus dipole, via the telescope mirror, into space; the curved field lines in the telescope aperture are shown resolved into the linearly polarized components that a distant observer would see. In (c), the right half of the telescope aperture is shown for the complementary situation: the straight field lines from a fully polarized distant source are shown resolved along the directions which the focus dipoles see as their orthogonal polarizations. (d) The cross-polarized aperture field pattern has 180° rotational symmetry. The cross-polarized antenna pattern is the Fourier transform of this and has similar structure; a real-life case is shown in (e). Polarization is generated from unpolarized incoming plane waves, with maximum conversion from directions along lines at 45° to the dipoles, roughly at the half-power points of the main co-polarized beam.

it is very stable and can be corrected for, after proper calibration by correlation methods (section 5.5.5).

5.5.5 *Polarization sidelobes*

Section 5.5.4 discussed the response of the polarimetric system to an unpolarized source located on the optical axis of the system. Radio systems also show considerable off-axis* 'polarization' responses to unpolarized radiation; such sidelobes are known as *(cross-)polarization sidelobes*. They are due to curvature of the electric field lines generated by the focal-plane antenna on the telescope mirror surface (engineers usually think – and talk – of antennas as transmitting into space; this is an example of the optical principle of reversibility; for receiving mode, one should think of the field lines which a distant source would have to produce on the mirror surface for the focal-plane antenna to see them as straight; fig. 5.4).

Design of focal-plane antennas for polarimetry concentrates on straightening the field lines in the telescope aperture and thus reducing the cross-polarization sidelobes. More details are given in Napier (1989, pp. 55–8), including a reference to what happens when the focal-plane antenna is not exactly on the optical axis of the telescope mirror(s) (which, of course, is a loss of symmetry, so one expects polarization effects on the effective axis of the complete telescope consisting of mirrors and focal-plane antenna; this is indeed what happens; see section 5.1, also Fiebig *et al.* (1991) and Schmidt *et al.* (1993)).

In synthesis instruments, the imperfections of the individual telescopes (such as the cross-polarization sidelobes and ellipticity of the primary beam) enter into the polarization response of the two-element interferometers, but the extent to which this happens depends on the relative orientation (crossed, parallel, $\pm 45°$) of the dipoles in the two telescopes. As noted above, the end results depend on the details of the synthesis installation concerned; at the time of writing, both insight and everyday practice are evolving.

A special case arises in Zeeman effect measurements at 21 and 18 cm, in an attempt to detect the very weak interstellar magnetic fields. The basic measurement is that of very faint polarized *spectral line structure*, and all continuum polarization is removed during measurement and reduction. We are therefore only concerned with polarizing sidelobes looking at spectral-line radiation (or, much less likely, continuum radiation through 'sidelobes with line structure' – which could possibly be a receiver artefact), and very low error levels are achieved. Even so, it is still doubtful whether any of the claimed magnetic field detections from Zeeman effect studies in radio *emission* lines are in fact reliable. Verschuur (1995a,b) addresses these questions (see fig. 5.5), and the discussion will no doubt continue for several years yet.

Since astronomical polarization signals are often small, low-level polarized sidelobe structures can contribute appreciable polarimetric errors when they happen to point at strong sources (the Sun, Cas A etc.). Since output varies

---

* i.e. in off-axis *directions*, even for an antenna mounted on the telescope optical axis (at the centre of the focal plane); see the note on p. 70.

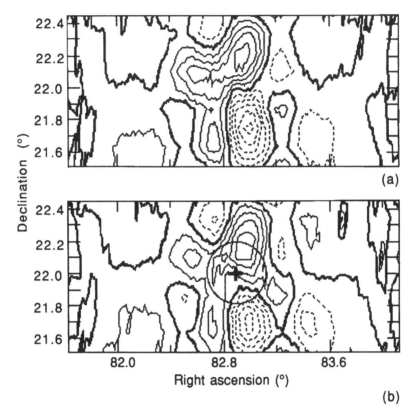

Fig. 5.5 Maps of 21 cm polarized beam structure obtained on Tau A using the NRAO 140 ft radio telescope, from Verschuur (1995b and private communication). (a) The average of six maps at feed rotation angle 155°. (b) The average of six maps at feed rotation angle 335°. The heavy contour is zero level, dotted lines are negative values and contours are at intervals of 0.05% of the main beam peak. The half-power beam is indicated by the circle in (b). Note that for a perfect feed antenna, these maps would be identical; the extent to which they differ indicates the residual uncertainties even when antenna patterns have been investigated thoroughly. Two components of the sidelobe structure (two-fold, four-fold symnmetry) are clearly identifiable; Verschuur suspects a third (which rotates with the feed antenna), at a 0.05% level.

linearly with source strength, a good first approximation to this sidelobe pattern can be obtained by scanning a strong source of low polarization through the antenna pattern. This method will always break down at *some* level, when competing signals from other sources through the main beam or through undocumented far-out sidelobes become dominant. The best method, therefore, is to use a correlation-type interferometer, using one telescope to track the strong source on-axis in one well-defined polarization, and to use the other telescope to scan the antenna pattern under test in some other polarization, while correlating these two signals in a standard correlator receiver. This eliminates most of the systematic errors; only those signal components that are correlated with the selected polarization from the strong source will contribute to the output. It is an extension of normal methods of obtaining a reliable sidelobe pattern of a radio telescope (Scott and Ryle 1977) and need not be discussed in detail here. Once the sidelobe structure has been determined reliably, and a first approximation to the polarized and unpolarized sky has been obtained, sidelobe contributions can be removed during data processing (iterative cleaning of the data). This approach is in its infancy as far as polarization is concerned and its limits are still unclear (but see Verschuur 1995a,b).

Troland and Heiles (1982a, figure 1 and discussion) and Wieringa *et al.* (1993, section 2.2) document far-out polarization sidelobes with a 'spokes' or 'cloverleaf' type of structure, which are probably caused by (polarized) diffraction around the support structure of the focal-plane assembly when the plane wave is on its way to the primary mirror; though the mechanism is different from the cross-polarization lobes, the measures taken to remove the errors from the data are similar. A complicating factor is that such far-out sidelobes may, for part of the measurement procedure, be looking at the local ground structure rather than at the sky; modelling radiation emitted by or reflected from such local ground structure is fraught with difficulties, but this situation can sometimes be approximated successfully; see section 5.5.6.

5.5.6 *Polarized radiation through sidelobes: ground reflections*

In radio-polarimetry of the galactic background, a strongly elevation-dependent, vertically polarized component was found to add to whatever sky polarization was detected (see Brouw and Spoelstra (1976) for a detailed investigation and earlier references; for those investigations, the elevation-dependent component was spurious and the term 'spurious radiation' has stuck; 'ground-reflection polarization' would be a more descriptive term). The primary mechanism is polarizing reflection of mostly unpolarized sky radiation by the ground surrounding the telescope; this reflected component then enters the system through the so-called spill-over sidelobe (the feed antenna looking over the rim of the main reflector with some residual, mostly non-polarizing, sensitivity). However, there must be other components contributing to the total. The only way to eliminate this polarization component has been to determine its average strength and (vector-)subtract that from all other measurements; variations in its strength remain in the observations as noise.

The assumption used when determining the average strength of this component is that truly celestial polarization will average out if one observes the sky at many different orientations. The Dwingeloo telescope, for which this is best documented, is alt-azimuth and the sky rotates with respect to it. At the start of any observing period, the practice has been to spend several days and nights taking one elevation scan after another, at many different azimuths; Q and U in the alt-azimuth system are then averaged per elevation and azimuth interval. See Brouw and Spoelstra (1976), and Spoelstra (1992), for further details.

Polarimetry by synthesis systems has been found to be less sensitive to this kind of error. This agrees with the mechanism described above: the correlated radiation from one point in the sky will, after reflection, enter the two telescopes of an interferometer with relative phase depending on local ground topology (detailed shape of the effective electrical earth plane); since many points in the sky contribute, the complex correlation will take on values which spread over the complex plane, and the average correlation will be very small. Sidelobes from direct-neighbour telescopes, however, can overlap on the ground, and, under these circumstances, 'spurious polarization' effects have been found in interferometers (T.A.Th. Spoelstra, private communication, 1994).

5.5.7 Conversion from one polarization form to another

Sections 5.5.4 to 5.5.6 have been concerned with spurious *polarizer* action of parts of the system. In this section we consider *retarder*-type actions, i.e. we disregard the first row and column of the Mueller matrix. An astronomically relevant example is partial conversion of strong optical linear polarization into circular polarization by the telescope. This will interfere with measurements of the minute interstellar *circular* polarization, which is thought to arise from weak birefringence of the interstellar medium acting on the strong linear polarization generated by the aligned dust grains further away from the observer. One has to consider the detailed circumstances when looking for a way to calibrate effects such as this; in the example discussed here, an alt-azimuth telescope effectively allows rotation of the telescope with respect to the sky, so that converted linear polarization can be made to change sign by 90° rotation (cf. the Mueller rotation matrix of section 4.1). For an equatorial telescope, the only point in the sky where one can accomplish rotation is the celestial pole (set the telescope to declination 90° and rotate it in hour angle). A good calibration source of strong linear polarization is daytime blue sky: measuring circular polarization while rotating the telescope allows one to disentangle true circular polarization from converted linear polarization; a separate measurement of the linear polarization, which is only slightly influenced by the very small circular polarization, then allows the conversion factor to be calculated with sufficient precision.

Similar effects occur in radio-polarimetry whenever components have not been adjusted properly, or at the limits of the passband, or far from the centre

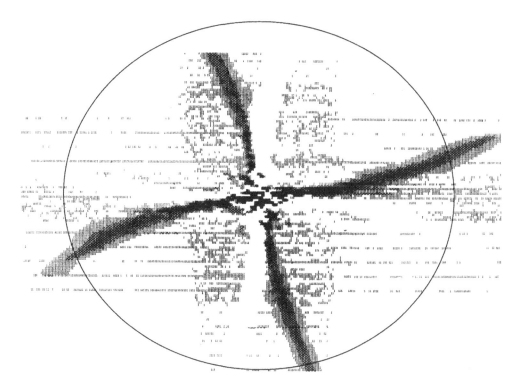

Fig. 5.6 Circularly polarized sidelobes out to 24° (the ellipse) from the optic axis (adapted from Troland and Heiles 1982a). Only the positive values are shown; the negative parts of the antenna pattern appear as blank areas.

of the synthesized field. Specialized examples can be found in Spoelstra (1992); here also, calibration is custom-designed for the case in point.

5.5.8 Errors in radio Zeeman polarimetry

Measurement of interstellar magnetic fields via the Zeeman effect for neutral hydrogen or the hydroxyl radical has been developed to a fine art (Heiles 1989, Verschuur 1989). The method is highly differential (polarization modulation and frequency switching). In the resulting spectral I profile, there are often minute remnants of errors due to equipment imperfections which do not quite cancel out. These resemble faint broad line components, and Gaussian component analysis reports them as such. When an attempt is made to trace them in the spectral V profile, ambiguous results are obtained. The *ad hoc* method adopted in practice is to discard these components, but some of them could, in fact, be real. The matter is discussed in some detail in Heiles (1989).

As was discussed in section 5.5.5, faint circularly polarizing sidelobes (fig. 5.6) can admit unpolarized line radiation masquerading as polarized radiation. All modern investigations make an attempt to eliminate such error contributions in the reduction stage (e.g. Verschuur 1989).

It is to be expected that detailed analysis will show that, in a synthesis instrument, some of the above error contributions can be excluded, leading to improved reliability of Zeeman polarimetry; synthesis instruments also have the inherent capability to determine the antenna pattern of the individual telescopes with great accuracy.

5.5.9 Ionospheric Faraday rotation

Faraday rotation in the Earth's ionosphere is variable and depends on the line of sight, but it is of order 1 radian at a wavelength of 1 m and is proportional to the square of the wavelength. Hence it is negligible for wavelengths shorter than about 10 cm, but increasing efforts must be made to eliminate it at longer wavelengths. Ionospheric sounder data are used to determine hourly values of certain parameters in a model for the local ionosphere; the exact form of the model and the exact choice of parameters depend on the geographic and geomagnetic latitudes of the observatory. The uncertain quantity is the 'total electron content' (vertical column density of electrons) of the ionosphere at its intersection with the line of sight; the Earth's magnetic field is sufficiently well known. Ionospheric sounders (Earth-based and topside) measure peak electron density (electrons per unit volume) and approximate layer thickness, from which the total electron content may be deduced. Much greater accuracy can be obtained by making use of satellite transmissions to measure the total electron content directly; the Global Positioning System (GPS) seems promising in this respect.

When the ionospheric Faraday rotation is sufficiently inhomogeneous within the telescope beam and/or receiver passband, depolarization will occur (section 4.1, example 5); this is only likely at long wavelengths, and will, in most cases, be negligible compared with Faraday depolarization within the astronomical source being observed.

5.6 Reduction of polarization observations

Reduction of polarimetric data requires a few additions to the standard repertoire of data-handling packages (arithmetic operators, scaling, rebinning). Like other observational data, most modern polarimetry is in the form of data arrays, but for polarimetry one often needs to handle two or more arrays in a single action, i.e. using vector manipulations. An often-required routine is conversion from Cartesian $(Q/I, U/I)$ to polar $(p, 2\theta)$ coordinates (and vice versa, although it is generally best to stay in a Cartesian system until final data presentation). Note the pitfalls in rebinning discussed in section 5.4.

The situation with respect to relevant documentation changes continually as newly developed instruments require new routines or adaptations of existing ones. Documentation should always be located via the user manual of the instrument used for the observations. For optical work, two or three versions of generally useful routines have been written and incorporated into IRAF (J. Walsh, adapted by R.G.M. Rutten), MIDAS (J. Walsh) and FIGARO (Bailey 1989; J. Walsh, adapted by R.G.M. Rutten). For single-dish radio-polarimetry, the NOD2 software system is available (Max-Planck-Institut für Radioastronomie, Bonn). For synthesis work, the VLA reduction package AIPS is best documented (but does require previous knowledge of synthesis theory). For Westerbork, a very effective polarization subsystem exists within NEWSTAR. The AIPS++ international synthesis reduction system, now being developed, is expected to have polarimetry fully integrated within it.

5.7 Polarization-induced errors in photometry

When a telescope or instrument is sensitive to the polarization of the input radiation, simple photometry of polarized objects will be in error (we are stating the obvious here, but how many people actually take notice?). In such cases, a complete polarimeter is needed to do accurate photometry and Mueller matrix analysis of the instrument will be required. The alternative (which in practice comes to nearly the same thing) is to depolarize the radiation before it reaches any polarization-sensitive part of the installation.

Optical examples of such systems are telescopes or instruments with oblique or off-axis mirrors and instruments with diffraction gratings (the so-called 'Wood's anomalies' in spectrographs are polarization errors in spectro-photometry). Relevant references are Tinbergen (1987a, 1988) and Murdin (1990); see also fig. 5.7.

In radio systems, polarization errors in photometry will occur if only one dipole is used in order to economize on electronics or on available correlator channels (e.g. to have four times as many correlators for simultaneous telescope spacings or simultaneous wavelength channels); this maximizes the polarization error, since a dipole is a near-perfect polarizer. It would be better to install both dipoles and to combine their signals with equal weight; this does involve a certain amount of top-end electronics, but could still allow economies on intermediate-frequency electronics and correlators. *If* there are good reasons to assume that celestial sources and all stray radiation are constant, the

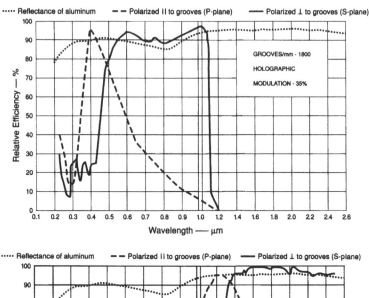

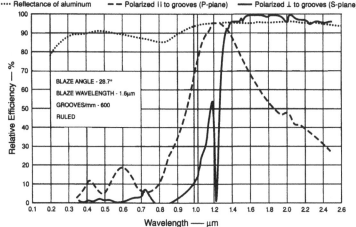

Fig. 5.7 Polarization reflectance curves of representative gratings, compared with the reflectance of a plane aluminized surface. Solid line: electric vibration perpendicular to the grooves of the grating ('S-polarization'); broken line: electric vibration parallel to the grooves ('P-polarization'); dotted line: aluminium (normal incidence, therefore any polarization). Charts and description courtesy of Milton Roy Company (Rochester, NY), a subsidiary of Sundstrand Corporation.

photometric maps in two orthogonal polarizations need not be obtained simultaneously. Needless to say, such 'good' reasons very often are no good at all.

In theoretical astrophysics, 'photometric' errors can result from the use of scalar radiative transfer theory when polarization effects are important and full Mueller matrix treatment should have been used. Mishchenko *et al.* (1994) investigated such errors in the context of Rayleigh-scattering atmospheres above a Lambertian surface. From that paper, from papers cited there and from Kattawar *et al.* (1976), the conclusion must be that 'photometric' errors up to 20% or even 30% are possible when polarization is neglected in computing the light reflected or transmitted by multiple-scattering atmospheres.

6

Instruments: implementations

This chapter will focus on those aspects of polarimetric instruments that are peculiar to certain wavelength regions. The concepts discussed in previous chapters will be used freely. *Non*-polarimetric wavelength-peculiar concepts will generally be taken for granted, but a few are essential and must be recapitulated briefly.

6.1 Optical/infrared systems

Optical polarimetric instrumentation has a long history of development. Early polarimeters had errors at the level of a few tenths of a per cent at best, and polarization signals were small, so that polarimetry was very much a specialist craft. B. Lyot was the first to obtain very high accuracy by devising a modulator and using it on the Sun. For stars, the signals were generally so small that photon shot noise was appreciable, and there was little incentive to design sophisticated systems of unavoidably smaller throughput.

The situation has changed drastically within the last decade or two. Larger telescopes are available, CCD detectors now offer thousands of parallel channels of potentially very good accuracy, and improved modulators of high transmission have been devised. The higher signal levels have meant that greater resolution (spectral, temporal, spatial) can be used, and this has had the effect of increasing the degree of polarization provided by nature (less smearing of polarizations from neighbouring resolution elements); the end result is that (i) many more situations within astronomy can usefully be tackled by polarimetry without exceptional cost in telescope time and (ii) 'common-user' polarimetry is becoming available in the optical/near-infrared wavelength region (the 'CCD domain'). The latest development is that such polarimetry is becoming available in the 1–5 μm region, where the detector arrays are improving fast and modulators similar to those at optical wavelengths can be constructed (see figs. 3.7 and 3.8

for published results); as in all infrared systems, much of the detailed design work is concerned with cooling as much of the system as possible, and the resultant equipment may *look* very different from its visible-wavelength counterpart.

All optical and near-infrared polarimeters can be analysed by Mueller matrix calculus (or Jones calculus when an interferometer is part of the system). It is sufficient to split the instrument into its simplest components, multiply all the matrices together and inspect the top row of the resultant matrix. If one expresses the response of a real-life detector (which depends on polarization of the light striking it) as that of an ideal detector (which responds only to I), preceded by a fictitious optical element which one includes in the matrix train, the top row of the matrix for the total optical train specifies the output I signal (and hence the detector output) for any input Stokes vector, and describes both the intended mode of the polarimeter and its errors. For an example, see Tinbergen (1973). If (as is usual in a modulation polarimeter) the polarization state of the light striking the detector does not vary with the state of the modulator, it is not necessary to know the details of the fictitious optical element within the detector. Those details enter only as a fixed gain factor, whereas we are interested in the quotient AC/DC, which represents the normalized Stokes parameters Q/I etc.; the fictitious optical element is therefore in general omitted, or included within an arbitrary gain constant.

6.1.1 Modulators

Polarization modulation is essential to accurate polarimetry in the optical spectral region. The technique used most often is to vary a retarder within the instrument. Radiation of both orthogonal polarizations passes through the same components for most of the instrument, and one strives to modulate only the polarization preference, leaving the Stokes I sensitivity constant; this generally means either switching a retarder from one state to another by some means, or rotating a constant retarder. Modulation may be included in the Mueller and Jones matrices of the system.

Two modulators are shown in fig. 6.1. They are examples only, but serve to illustrate the basic principles. The general form of their Mueller matrices, including the analysing polarizer but excluding component 1, is:

$$0.5 \times \begin{pmatrix} A & B & C & D \\ A & B & C & D \\ 0 & 0 & 0 & 0 \\ 0 & 0 & 0 & 0 \end{pmatrix}$$

For the circular modulator

$$D/A = f(t), \quad |\text{Alt}(B \text{ and/or } C)/A| \approx \epsilon, \quad |\text{Alt}(A)/A| \approx \epsilon^2$$

while for the linear modulator

$$(B \text{ and/or } C)/A = f(t), \quad |\text{Alt}(D)/A \approx| \epsilon, \quad |\text{Alt}(A)/A| \approx \epsilon^2$$

where $f(t)$ represents an alternating function of time with amplitude close to unity, $0.5A$ is the transmission for unpolarized light, 'Alt(x)' is short for 'the part of x that alternates with the same frequency as $f(t)$', and ϵ is a small quantity (preferably of order 1%). The modulating function $f(t)$ is a square wave for the circular modulator and a sine wave for the linear one; including component 1 of fig. 6.1 in the instrument would reverse this and also exchange the values of B (and/or C) and D.

In the optical region, the only good polarizers available are linear polarizers (commercial 'circular polarizers' are always a combination of a linear polarizer with a quarterwave plate).* Until the beam passes through the analyser, its Stokes I is constant and all that happens is that *the polarization form* of the polarized part of the beam is switched between the two eigenmodes of the analyser (which in this case transmits one, absorbing the other, like a Polaroid). The function of the analyser is to convert the modulation of the polarization into modulation of Stokes I, which can be detected reliably by standard electronics.

Component 2 is the actual modulator. As the circular modulator rotates, the halfwave sections reverse the sense of circular polarization; this is a '+/−' type of modulation, leading to a square wave of known phase in the output of the analyser. In the linear modulator, the direction of vibration beyond the rotating halfwave plate rotates as well (twice as fast as the halfwave plate), leading to a sine wave in the analyser output; the phase of this sine wave corresponds to the polarization angle at the input. In both cases, the size of the modulated Stokes I component in the output beam, as a fraction of the total Stokes I signal, corresponds to the degree of polarization of the input light.

Fig. 6.4 shows a modern rotating-wave-plate polarimeter. The sine-wave modulation is rather slow, and scintillation- and extinction-noise will show through, but, for relatively faint objects with high polarization (fig. 3.3), photon noise will dominate and no deterioration will result from the slow modulation rate.

---

* Note added in proof: This situation may change in the near future through modern liquid-crystal developments (Philips press release 1996).

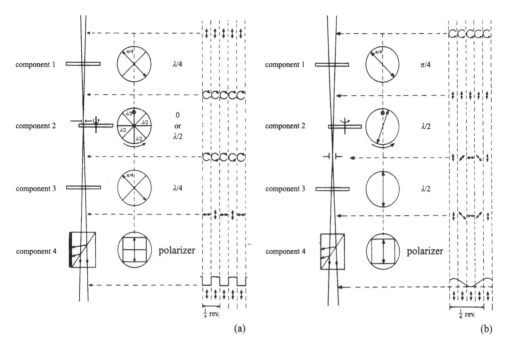

Fig. 6.1 Two polarization modulators: (a) for circular polarization, (b) for linear polarization. Both use a rotating halfwave plate as the modulating element; the basic modulator in each case comprises only components 2 to 4, while component 1 is there to allow the modulator to be used for both circular and linear polarization (Tinbergen 1972, 1974). Component 3 for the circular modulator is essential; it converts the circular polarization of the modulator to the linear polarization suitable for the analyser. For the linear modulator, component 3 is there to correct certain imperfections of the modulating halfwave plate.

Rotating components generally limit modulation frequencies to about 100 Hz, which is insufficient to suppress all scintillation noise. 'Electro-optic crystals' (or 'Pockels cells') can modulate faster, as can 'photo-elastic' (or 'stress-birefringence') modulators. The former type consists of a crystal that changes its birefringence when an electric voltage is applied to it and is generally operated as a square-wave modulator. The latter type (Kemp 1969) is a piece of glass (or fused silica for ultraviolet transmission) in mechanicallly resonant oscillation and thus with time-varying stress-birefringence (fig. 6.2). Since the last component is a linear polarizer, the Mueller matrix of the photo-elastic modulator has the same general form as before; in this case

$$A \approx 1$$
$$B = 0$$
$$C = C_0 + \text{even harmonics of the (mechanical) resonance frequency}$$
$$D = \text{fundamental and odd harmonics of the resonance frequency}$$

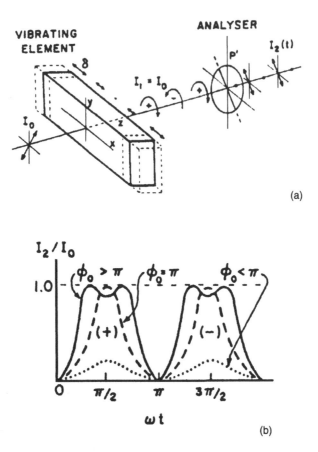

Fig. 6.2 The photo-elastic modulator, adapted from Kemp (1969). The following is Kemp's own description: (a) 'The birefringence modulator in rudimentary form. A simple extensional vibration is set up in a transparent bar, sustained by an acoustic transducer (not shown).' (b) 'The output flux I_2 as a function of time, for 100% linearly polarized input. The (+) and (−) peaks correspond to opposite elliptic or circular polarizations in the beam incident on the analyser.' ϕ_0 is the retardation amplitude at the centre of the oscillating bar. For more detail, Kemp's paper should be consulted.

Tuned amplification or synchronous demodulation in the electronics is then used to select the desired periodic term (generally, but not always, the funda-mental). The photo-elastic modulator is the most nearly perfect polarization modulator known; its maximum birefringence is only about one part in 40 000, so the oscillations do not influence the optical path of the light beam in any sig-nificant way (and therefore there is no significant spurious periodic signal which could be interpreted as being polarized); it can also be perfectly transparent over a wide range of wavelengths and will tolerate large angular width of the beam (Nordsieck *et al.* (1994b) report a 'DC' application of stress-birefringence

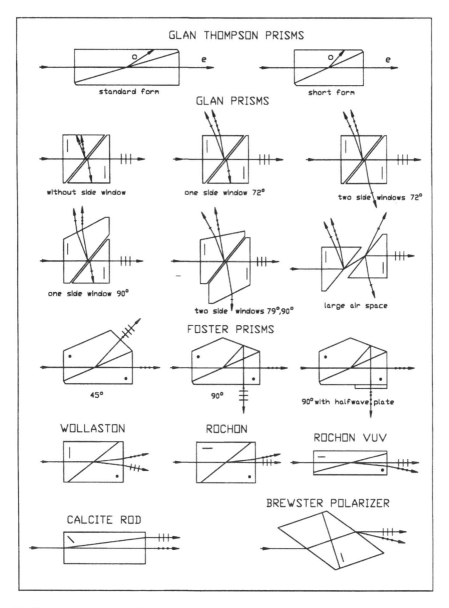

Fig. 6.3 The more common prism polarizers (courtesy of Bernhard Halle Nachfl. GmbH, Berlin). The choice between them is a matter of engineering design (one or two beams, location at pupil or image, transmission, stray reflections, ray geometry, component size and cost, etc.).

which exploits these properties for a wide-field polarimetric survey camera in the far-ultraviolet). The only flaw of the photo-elastic modulator is that it cannot easily be constructed in achromatic form (however, it can readily be tuned in wavelength).

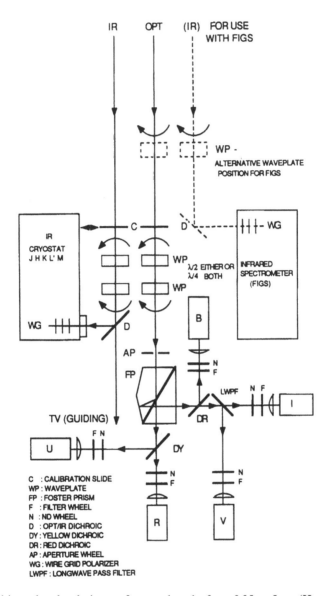

Fig. 6.4 A multi-passband polarimeter for wavelengths from 0.35 to 5 μm (Hough *et al.* 1991). Although a two-beam analyser is used, each band uses only one of the beams. A modulation period of 0.8 s is used. The instrument is optimized for time-resolved polarimetry within a wide wavelength range.

More complex modulators, for 'full Stokes polarimetry' (i.e. simultaneous determination of all four Stokes parameters), are required for solar and specialized stellar work; they are reviewed by Stenflo (1984 and 1994, sections ·13.7.2–13.7.4).

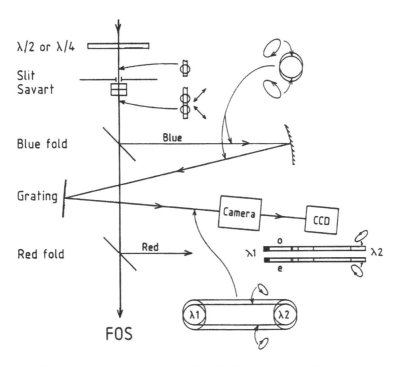

Fig. 6.5 The ISIS spectro-polarimeter of the 4.2 m William Herschel Telescope on La Palma; the red and blue fold mirrors can be implemented as dichroic or other beamsplitters, allowing simultaneous spectro-polarimetry in three channels at some sacrifice of flux efficiency. The wave plate (halfwave and/or quarterwave) can be rotated, thus serving as a polarization modulator; the analyser (Savart plate) has two exit beams with orthogonal polarizations, and the detectors are CCDs; beam cross-sections and polarizations are sketched at several positions within the optical system. Further details can be found in Tinbergen and Rutten (1992); see also Schmidt *et al.* (1992b) for a system designed on similar principles, but, in actual fact, different in almost every detail.

6.1.2 Two-beam analysers

Detectors such as CCDs generally require a minimum integration time of the order of seconds, so that fast modulation cannot be used. Stepped rotation of a wave plate, with a separate integration at each position, can be regarded as a very slow modulation. The unavoidable extra noise in such a system can be eliminated if the analyser is of two-beam construction and both beams are recorded: 'common-mode' noise, such as from scintillation and other atmospheric variations, affects both beams equally, while the modulator is used to label truly polarized light by switching it from one beam to the other for successive exposures. Two-beam polarizers exist in many forms (fig. 6.3); the Wollaston prism and the calcite plate ('calcite rod' in fig. 6.3) are used most in astronomy. At (sub-)millimetre wavelengths, wire-grid polarizers can be used (Hildebrand *et al.* 1984, Clemens *et al.* 1990).

The following extract from the Users' Manual for the ISIS spectro-polarimeter (fig. 6.5) explains how such very slow 'modulation' is used:

*The polarimeter uses a Savart calcite plate, which yields 2 spectra (of opposite polar-
ization). The polarization information (one Stokes parameter per exposure) is contained
in the ratio, at each wavelength, of the intensities in the 2 spectra but it is mixed up
with the system gain ratio for the pixels concerned. The effect of the unknown gain is
eliminated by inverting the sign of the polarization effects in a second exposure, while
leaving the gain ratios identical. Inversion of (linear) polarization effects and therefore
of the Stokes parameters is accomplished by rotating the halfwave plate by 45 degrees;
while the polarization effects are inverted, the system gains remain the same since these
are determined by the built-in polarization of the o and e exit beams of the calcite plate.
All instrumental conditions (grating parameters, filters, dichroics, Dekker, slit, etc.) must
be the same in both exposures; image centering on the slit is the most difficult to control
in this respect.*

*The derivation of Stokes parameters from the recorded spectra is presented below. We
factorize the conversion 'constant' for input light to detector signal into a polarization-
dependent, time-independent part G and a time-dependent, polarization-independent part
F: $G_{\parallel}$ and $G_{\perp}$ refer to the o- and e-spectra on a single frame; they include grating
efficiencies and reflection coefficients of mirrors, and the sensitivity of the pixel considered
to the polarized light striking it. F_0 and F_{45} refer to the two separate frames (halfwave at
$0°$ and $45°$) and include atmospheric transmission, seeing, image wander and variations
in shutter timing. I and Q refer to total and polarized light input and the i refer to signals
recorded by the detector. $p_Q = Q/I$ is the Q component of the degree of polarization.*

$$i_{0,\parallel} = 0.5(I + Q) \cdot G_{\parallel} \cdot F_0$$
$$i_{0,\perp} = 0.5(I - Q) \cdot G_{\perp} \cdot F_0$$

$$i_{45,\parallel} = 0.5(I - Q) \cdot G_{\parallel} \cdot F_{45}$$
$$i_{45,\perp} = 0.5(I + Q) \cdot G_{\perp} \cdot F_{45}$$

*To derive Stokes parameters from these spectra, first divide the o- and e-ray spectra in
each frame to take out the scaling factors F. Dividing these ratios again cancels the G
factors. The Q Stokes parameter, in degree-of-polarization scale, is:*

$$p_Q = \frac{R - 1}{R + 1} \qquad \text{with} \qquad R^2 = \frac{i_{0,\parallel}/i_{0,\perp}}{i_{45,\parallel}/i_{45,\perp}}$$

*Note that by multiplying the intermediate ratios, instead of dividing them, the G ratio
(relative flat field) is obtained. The other Stokes parameter, p_U, is obtained similarly
from the pair of exposures with the halfwave plate at $22.5°$ and $67.5°$. The raw degree
of polarization p and polarization angle θ are then given by:*

$$p = \sqrt{(p_Q)^2 + (p_U)^2} \qquad \text{and} \qquad \theta = 0.5 \arctan(p_U/p_Q)$$

The equations for $I + Q$ etc. above can be understood in terms of Mueller
matrices as follows: The matrix for a halfwave plate in positions $0°$ (upper

sign) and 45° is:

$$
\begin{pmatrix}
1 & 0 & 0 & 0 \\
0 & \pm1 & 0 & 0 \\
0 & 0 & \mp1 & 0 \\
0 & 0 & 0 & -1
\end{pmatrix}
$$

Multiplying this by the matrix for a polarizer at 0°, we obtain, for the $+Q$ polarimeter

$$
0.5 \times
\begin{pmatrix}
1 & \pm1 & 0 & 0 \\
1 & \pm1 & 0 & 0 \\
0 & 0 & 0 & 0 \\
0 & 0 & 0 & 0
\end{pmatrix}
$$

For the other beam from the analyser ('Savart' in fig. 6.5), the output polarization is at 90° and the resultant matrix of the $(-Q)$ polarimeter becomes

$$
0.5 \times
\begin{pmatrix}
1 & \mp1 & 0 & 0 \\
-1 & \pm1 & 0 & 0 \\
0 & 0 & 0 & 0 \\
0 & 0 & 0 & 0
\end{pmatrix}
$$

Using the definitions of the F, G and i in the extract above, the equations follow directly from the top rows of these two matrices. Note that any circular polarization in the input does not influence the output I in either of the beams (provided the retarder is indeed exactly halfwave). The data reduction outlined shows that scintillation noise is eliminated by removal of the F from the final result. Another way to understand this is as two separate modulation polarimeters, for $+Q$ and $-Q$. Both are affected by the same scintillation and extinction noise and, in the process of obtaining the average of $+Q$, this coherent component of the noise is inverted for the $-Q$ polarimeter.

We could of course have rotated the halfwave plate to a large number of positions and analysed for the characteristic sine-wave pattern; but the modulation would have been very slow indeed. Having verified in checkout procedures that a true sine wave is in fact obtained, the preceding method is the least redundant way of finding the linear Stokes parameters, by modulation reduced to its minimal form, '$+/-$' modulation of Q or U.

There are two ways of looking at this process: one is to say that we convert the input linear polarization to its orthogonal form by rotating the halfwave plate by 45°, the difference in output of the polarimeter reflecting the switched input polarization; the other is the view taken above, viz. to include the halfwave plate within the polarimeter and to say we have switched the 'preference' of the polarimeter from one polarization to its orthogonal form.

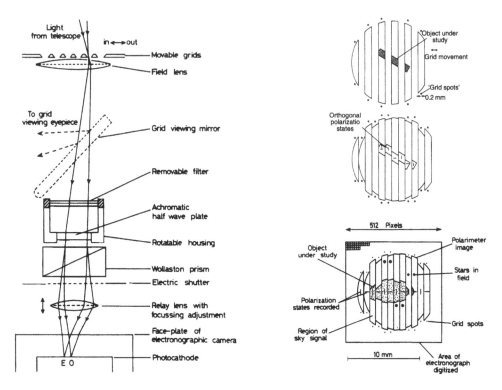

Fig. 6.6 The Durham imaging polarimeter, from Scarrott *et al.* (1983). The original detector of this polarimeter was electronographic; a CCD has now replaced that system. This compact travelling polarimeter has seen a great deal of service on telescopes all over the world.

The essential part of either view is that one changes *nothing whatsoever* in the instrument, except to rotate the halfwave plate to a new position angle and to take a second exposure with this orthogonal state of the polarimeter. *As long as nothing is changed,* it does not matter in the least that the grating and folding mirrors respond differently to different polarizations; all this is included in the G values, which were assumed constant with time (the extent to which the assumption of factorization breaks down will determine the ultimate error level for this method).

The imaging polarimeter of fig. 6.6 uses a Wollaston prism as analyser, but is otherwise similar to that in the spectrograph of fig. 6.5. Since the image in this case is two-dimensional, half of it is blocked for any one exposure, and a total of eight exposures is needed for full linear polarimetry of the entire input image. A Zeeman polarimeter for solar speckle-and-restoration imaging is described by Keller and von der Lühe (1992); no modulator was used in this test system, but a complex modulator for full Stokes polarimetry is envisaged.

A variant of the two-beam analyser polarimeter uses a chopper and a single (photomultiplier) detector to record the two beams alternately (see Piirola (1973) and, in a simultaneous five-colour version, Korhonen *et al.* (1984)), a practice analogous to the use of the phase switch in the radio-polarimeter of fig. 6.12. In principle, the method may be combined with a rotatable halfwave plate to yield even higher freedom from systematic error. Until fast-readout CCDs are integrated into polarimetry (Tinbergen 1995), this dual-beam chopper approach yields a unique combination of wide spectral band and very high polarimetric accuracy. On an alt-azimuth telescope (section 5.5.4), such a polarimeter is the ideal instrument for fundamental establishment of a linear zero-polarization standard in the sky, through very systematic observations of bright stars of low polarization.

6.1.3 Achromatic systems

We have so far used polarizers and retarders in conceptual polarimetric instruments without worrying about how one actually constructs them. In a typical astronomical application, we need data at many wavelengths and, to make the best use of the little light we have, we require our instruments to work over the widest possible wavelength range. In practice, this means that, in the optical region, we should like to work with a single instrument from $0.3\,\mu$m to about $5\,\mu$m. Existing detectors cannot span this range, so that separate instrumentation is usually built for wavelengths below and above about $1\,\mu$m; polarization components should preferably perform at all wavelengths within these ranges; in other words, they should be achromatic to a high degree.

Polarizers are most readily available in achromatic form, in the sense that they do separate the two beams effectively for a wide wavelength range. The amount of separation (in position or angle) is often a function of wavelength, but the design of the rest of the instrument can usually accommodate that. (For example, in the ISIS spectro-polarimeter of fig. 6.5, the separation of the two star images on the slit depends slightly on wavelength, which means that the spectra on the CCD are not exactly parallel, nor straight; this does not matter much, as long as all the signal in each spectrum is gathered during reduction, the sensitivity of the CCD is more or less constant over its surface and the images do not overlap; cf. Schmidt *et al.* (1992b, section 2.1).)

In nature, most retarders are very *chromatic*: their retardation varies strongly with wavelength. The most common type of retarder is a slice of crystal, of which the retardation δ takes the form:

$$\delta = 2\pi x(n' - n'')/\lambda$$

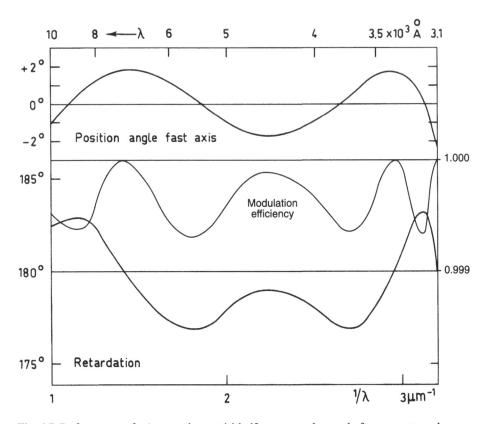

Fig. 6.7 Performance of a 'superachromatic' halfwave retarder made from quartz and magnesium fluoride; adapted from Tinbergen (1974). The polarimetric modulation efficiency exceeds 0.999 over a 3 to 1 range in wavelength; such a halfwave retarder can also serve as a very good 'linear depolarizer' (q.v.).

$n' - n''$ is a slow function of λ, so that δ is approximately proportional to $1/\lambda$, a very unsatisfactory situation.

One way to achromatize retarders is to use two materials with different wavelength-dependence of birefringence $n' - n''$ and to use a negative multi-wave plate of one material to cancel all but a small fraction (half- or quarter-wave) of a positive multi-wave plate of another material (fast axes of the two components crossed). By proper choice of the thicknesses, one may thus implement the desired retardation at two wavelengths, with a roughly parabolic variation through the rest of the spectrum.

With birefringent polarization components, there is another option, viz. to combine three or more slices *of the same material* into an achromatic combination (e.g. Pancharatnam (1955a,b); see also fig. 4.1 and table 4.2, and exercises 4.10 and 4.11). Since the only requirement is that the relative wavelength dependence is the same for all the constituent slices, these may themselves be

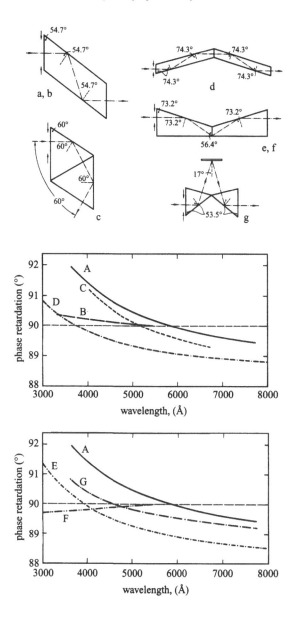

Fig. 6.8 The Fresnel rhomb (top left) and alternative total-internal-reflection retarders, from Bennett and Bennett (1978), reprinted with permission of McGraw-Hill, New York. The fast axis direction of these retarders is determined by the geometry, and is independent of wavelength. Spectral performance curves (capitals) refer to the components (lower-case) shown. Curves B and F refer to optimized versions of components a and e, with slightly changed angles of reflection and a thin magnesium fluoride coating applied to the internal-reflection face.

two-material achromats. Such a doubly achromatized combination is known as *superachromatic*; it is difficult to construct (at least six slices of rather precise thickness of a few tenths of a millimetre each, oriented precisely and cemented in place without damaging the fragile crystals) and is very expensive, but can yield > 99% modulation efficiency from 300 nm to 1 μm or beyond (fig. 6.7).

The achromatization procedure with three slices of the same material was devised by Pancharatnam (1955a,b); similar procedures exist for achromatic polarization rotators (Koester 1959) and for larger numbers of constituent plates, but none of these are used very much. The Pancharatnam combinations show dispersion of the axis directions, i.e. the position angle of the fast (and slow) axis is a function of wavelength. This means that the phase of the sine modulation by a rotating halfwave plate is a function of wavelength (e.g. in the superachromatic version, which is of Pancharatnam construction; see fig. 6.7). This does not matter in a spectrograph, since it can easily be calibrated for each wavelength; however, it may be troublesome if it is used with a wide-band filter (e.g. in an imager for faint object polarization). For the rotating-halfwave-plate modulator, the problem can be eliminated by inserting a second, identical but (usually) stationary, halfwave plate in the beam (component 3 in fig. 6.1b).

Another type of achromatic retarder makes use of total internal reflection within a prism. Such reflection introduces a phase difference which is a function of angle of incidence and refractive index; however, this functional dependence is slow enough for the component to be very useful whenever a straight-through beam is not required. The 'Fresnel rhomb' (fig. 6.8) is one of the retarders of this type, and produces about 90° retardation in two reflections. Two Fresnel rhombs in series can return the beam to its original path and provide very achromatic halfwave retardation; the only disadvantages of such a combination are its length and the delicate mechanical adjustments necessary for its operation.

More detailed descriptions of achromatic components may be found in Tinbergen (1973, 1974) and in Bennett and Bennett (1978). The instruments in figs. 6.4, 6.5 and 6.6 all use a superachromatic halfwave plate.

6.2 Radio systems

We have seen that, in optical polarimeters, orthogonal polarization signals follow the same path through the optical system. Birefringent components are used to generate a distinction between these two signals, but the same system gain applies to both of them separately, so that, apart from the system gain as a multiplying factor, the difference between them is reproduced faithfully. The gain can be calibrated reliably by observing a polarized source. Radio receiver channels are normally not birefringent and they lack the capability to process a pair of polarization signals within one physical channel; two channels are required, and these channels have different gains, the ratio of which inevitably varies somewhat with time (and in astronomy we are particularly sensitive to this, since we usually have a small polarization difference between two large signals, each representing about half the total radiation). One could calibrate the time-varying ratio, but this would only transfer the problem to the stability of the calibration arrangements (both the source and the coupling circuitry). Rather than determining a difference or ratio of two large and spuriously varying intensity signals, in radio-polarimetry one determines the polarized component directly, by measuring the correlation between the electric fields of orthogonal polarization forms. This is possible because in radio receivers one has available an electrical signal which represents wave *amplitude and phase* rather than 'intensity'. The correlation technique is introduced in section 6.2.2.

Polarization processing is carried out mainly at *intermediate frequency*; phase-maintaining conversion to intermediate frequency is a technique which in astronomy is peculiar to the radio domain, and it is introduced briefly in section 6.2.1.

Note: Another peculiarity of radio techniques is that the photons are of such low energy that even the smallest signals contain very many of them and photon shot noise is vastly exceeded by the noise of the first stage of electronic processing. This means that after amplification one may divide a radio signal into a number of identical copies and process these independently without worrying about introducing extra photon noise by having split the energy of the signal over several channels. One has indeed split the signal, and therefore fewer photons are available in each channel, but the effect of this on total noise is negligible. In synthesis instruments, one correlates *simultaneously* the output of each (telescope/polarization) channel with all other channels, and achieves a gain in observing speed by this massive parallellism. In contrast, if one were to divide an optical signal into many parts, one would increase the fractional noise in each channel and there would be no net gain in total speed; this is because the best optical detectors have very low noise compared with that of the energetic optical photons, and dominant noise in most optical systems is therefore 'noise-within-signal' (photon shot noise) rather than amplifier or detector noise (however, CCDs have only recently attained 'one-electron' readout noise, and infrared arrays do have significant readout noise).

6.2.1 Frequency conversion in a nutshell

Frequency conversion is a technique whereby a signal is transferred to a (generally lower) carrier frequency *with retention of relative amplitudes and phases*; the passband and the signal components within it are faithful replicas of the originals at the higher carrier frequency. Since some of the polarization

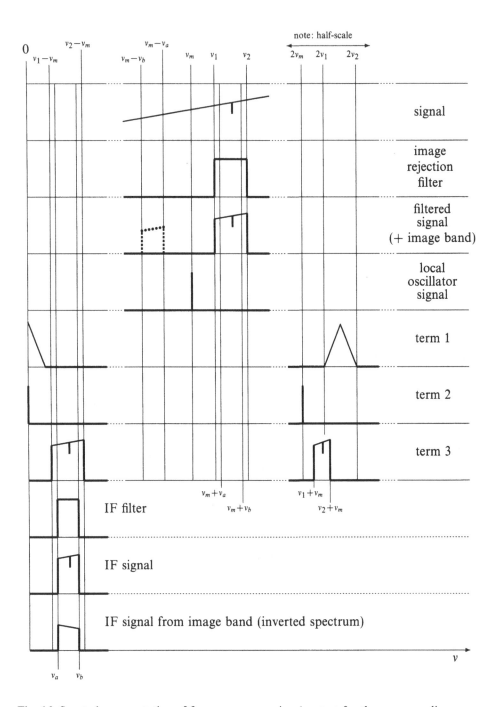

Fig. 6.9 Spectral representation of frequency conversion (see text for the corresponding formulas); the lowest line shows how (the mirror image of) an undesirable part of the spectrum can also get into the output, unless an 'image rejection filter' is fitted to exclude that part of the signal before it is converted.

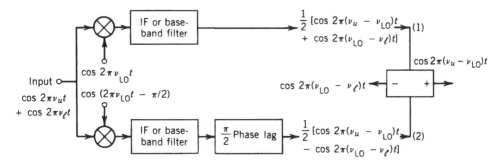

Fig. 6.10 An image-frequency-rejecting mixer (adapted from Thompson *et al.* 1986). The subscripts u and l refer to the upper and lower sidebands, respectively. The upper-sideband signal is obtained from the sum of the outputs (1) and (2), the lower-sideband signal from their difference.

processing in radio systems takes place at the lower, 'intermediate' frequency (IF), a summary of the technique is desirable.

In the *mixer*, a signal at LOCAL OSCILLATOR FREQUENCY is added to the input signal (SIGNAL FREQUENCY) and the sum of the signals is squared. The squared signal is then filtered to retain only a band of frequencies around the INTERMEDIATE FREQUENCY. See fig. 6.9 for a graphical representation; in simple equations:

$$\text{signal}: \quad A = \sum_{\nu_1}^{\nu_2} a_i \cos(2\pi \nu_i t + \phi_i)$$

$$\text{local oscillator}: \quad B = b \cos(2\pi \nu_m t + \phi_m)$$

The mixer forms

$$(A+B)^2 = A^2 + B^2 + 2AB$$

$$= \sum_{\nu_i = \nu_1}^{\nu_2} \sum_{\nu_j = \nu_1}^{\nu_2} a_i a_j \cos(2\pi \nu_i t + \phi_i) \cos(2\pi \nu_j t + \phi_j) \tag{1}$$

$$+ b^2 \cos^2(2\pi \nu_m t + \phi_m) \tag{2}$$

$$+ 2b \sum_{\nu_i = \nu_1}^{\nu_2} a_i \cos(2\pi \nu_i t + \phi_i) \cos(2\pi \nu_m t + \phi_m) \tag{3}$$

Using $2 \cos x \cos y = \cos(x+y) + \cos(x-y)$ and $2 \cos^2 x = 1 + \cos 2x$, we conclude that

- term (1) contains frequencies between $2\nu_1$ and $2\nu_2$ *and* between $\nu_1 - \nu_2$ and $\nu_2 - \nu_1$;
- term (2) contains the frequencies $2\nu_m$ and 0;

• term (3) contains frequencies between v_1+v_m and v_2+v_m *and* between v_1-v_m and v_2-v_m.

The IF filter passes only (part of) the latter of these frequency intervals, and finally we retain just:

$$b \sum_{v_i=v_m+v_a}^{v_m+v_b} a_i \cos[2\pi(v_i - v_m)t + \phi_i - \phi_m]$$

which is a true copy of the input signal at a lower carrier frequency. We therefore have the choice of manipulating at signal frequency or at intermediate frequency. Correlating two signals is simplest at intermediate frequency, while polarization discrimination by the dipoles is at signal frequency. Manipulating relative phases of signals is best done by manipulating the phase of the local oscillator, ϕ_m; the effect of manipulating the phase of a single frequency is then identical over the whole passband (achromatism).

Fig. 6.9 includes an 'image frequency filter'. For signal-to-noise reasons, such a filter is sometimes dispensed with, both signal and image frequencies appearing in the output; operation is then said to be 'dual-sideband'. In polarimetry, this can be a risky procedure when Faraday rotation is present or suspected. So-called single-sideband or image rejecting mixers (fig. 6.10) can then be used to make the two sidebands available separately (see Thompson *et al.* (1986, pp. 207–8); some polarization receivers actually require sideband separation to function properly (McKinnon 1992b)).

6.2.2 *Correlator polarimetry*

A 'correlator' is a standard component in radio astronomy; aperture synthesis depends on correlation-type interferometry. Basically, a correlator multiplies the instantaneous voltages in two receivers. It can be implemented in analogue form by squaring both sum and difference of the two signals and taking the difference of these two squares. It can also be implemented in digital form, in which case the numbers representing the quantities to be multiplied are quantized to 1 (sign) bit, or a very few bits at most. Such extreme quantization does mean a 10% to 20% loss in signal-to-noise ratio, but the gain in stability and simplicity of construction is immense. To obtain simultaneous spatial and spectral resolution, thousands of such few-bit correlators are used in large synthesis installations like the WSRT and the VLA, with system efficiency limited mainly by available funds. Discussion of digital techniques is beyond the scope of this book; the reader is referred to Thompson *et al.* (1986, section 8.3).

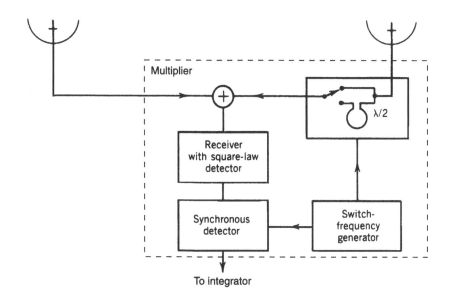

Fig. 6.11. A phase-switching multiplier, adapted from Thompson *et al.* (1986).

Being a multiplier, a correlator in radio astronomy is similar to the mixer used in frequency conversion, with the local oscillator signal replaced by a second signal of the same frequency content as the first and the resulting intermediate frequency equal to zero. The A^2 and B^2 terms are eliminated by phase switching (fig. 6.11). Fig. 6.12 shows how, for linearly polarized feed antennas, this can be used to measure linear polarization, with rotation of the feed antenna to obtain both Q and U. Unpolarized, and in this case circularly polarized, signals do not lead to an output in the correlator channel.

The feed antenna system can be made sensitive to other polarization forms, in particular to circular polarization. Fig. 6.13 shows how elliptically polarized feed antennas can be conceived and actually constructed as a network linking two linearly polarized antennas. Seen as a transmission system fed from terminal A, this is a hardware implementation of chapter 2, where all polarization forms were conceptually generated from two correlated linear polarization signals in certain proportions and with a 90° phase difference between them. As a receiving system, it produces an output at A whenever the dipoles receive correlated signals which correspond to the polarization form selected. When $\beta = \pm\pi/4$, the system is adjusted for transmission (and reception) of circularly polarized radiation; $\beta = 0$ or $\beta = \pi/2$ denotes adjustment for linear polarization.

Fig. 6.14 shows a more detailed schematic of a practical polarization receiver

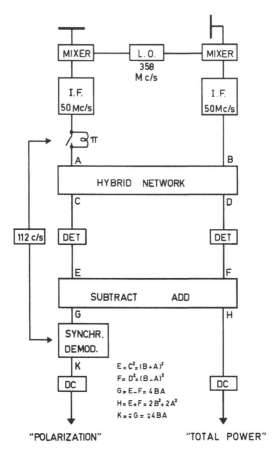

Fig. 6.12 A simple (linear) radio-polarimeter, from Westerhout *et al.* (1962). If the voltages at A and B are represented by the sum of a linearly polarized part (subscript 0) and an unpolarized part (subscript 1), as follows:

$$A = \pm\{e_0 \sin(\omega t + \alpha_0) \cos \chi_0 + e_1 \sin(\omega t + \alpha_1) \cos \chi_1\},$$
$$B = \{e_0 \sin(\omega t + \alpha_0) \sin \chi_0 + e_1 \sin(\omega t + \alpha_1) \sin \chi_1\},$$

then the smoothed 'polarization' output signal is proportional to $-I_0 \sin 2\chi_0$, while the smoothed 'total power' output signal is proportional to $I_0 + I_1$ (I representing $e^2/2$ and all the $\sin \chi_1$ and $\cos \chi_1$ terms having averaged to zero). Rotating the dipole focus antennas, one may derive the polarization angle χ_0, the linearly polarized signal I_0 and the degree of linear polarization $I_0/(I_0 + I_1)$.

with linearly polarized feed antenna, while fig. 6.15 shows a system using a circularly polarized feed antenna. These systems are described in Berkhuijsen *et al.* (1964) and Turlo *et al.* (1985),[*] respectively. The systems shown in

---

[*] The authors discuss questions of calibration, a perennial problem in radio techniques (electronic circuits are not as stable as pieces of optics). Readers should be aware that the authors assume $V = 0$ and use a Stokes 3-vector and Mueller 3×3 matrices; they also use an abbreviated and slightly confusing notation as follows: 'A/B' means 'A correlated with B' or 'A multiplied by B', 'COS' means that the inputs to the correlator are direct, 'SIN' means that one of them has had its phase shifted by 90°. Nevertheless, the paper is worth reading in detail.

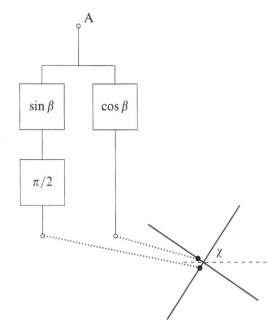

Fig. 6.13 A focal-plane feed antenna for polarimetry, after Thompson *et al.* (1986). The system may be made sensitive to any elliptical polarization by choosing values of χ (position angle) and β (arctangent of ellipse axial ratio); 'cosβ' and 'sinβ' stand for the voltage reponses of the units shown, and the unit labelled '$\pi/2$' is a phase shifter (the equivalent of the quarterwave plate of the optical domain).

figs. 6.12, 6.14 and 6.15 illustrate the evolution of radio-polarimeters from simple beginnings based on first principles to more complex modern systems, which, by specialized design, reduce the effects of the instabilities that are the curse of analogue electronics; it is clear that the practical design issues are of a different kind from those in the optical region. Complexity increases by another order of magnitude when one proceeds to the aperture synthesis systems described in section 6.2.3.

In dual-sideband receivers, special measures must be taken to ensure proper handling of correlator polarimetry (e.g. McKinnon 1992b)).

6.2.3 Polarimetry by synthesis arrays and VLBI

Aperture synthesis and VLBI systems use correlators in one form or another to determine the correlated part in the signals received by different telescopes. Such correlations provide information about the *angular* distribution of radiation over the sky, since they are values of the *spatial* auto-correlation function and the relation of that to the angular power distribution in the sky is a Fourier transformation.

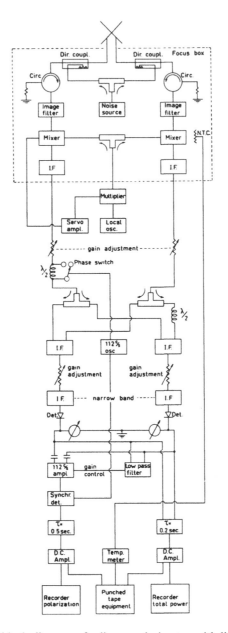

Fig. 6.14 More detailed block diagram of a linear polarimeter, with linearly polarized feed; details in Berkhuijsen *et al.* (1964).

It is a relatively simple extension, in principle, to cross-correlate (in four combinations) two orthogonal polarizations of the field received by each of two different telescopes and thus to determine values of the spatial auto-correlation function of Stokes Q, U, or V as well as I (cf. the second – i.e. $E_x E_y^*$, etc. – of the Stokes parameter definitions in section 4.3; the 'correlated'

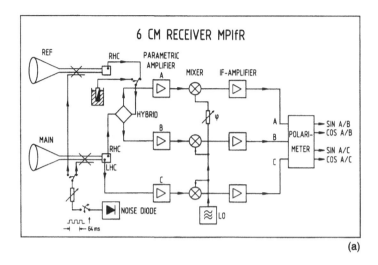

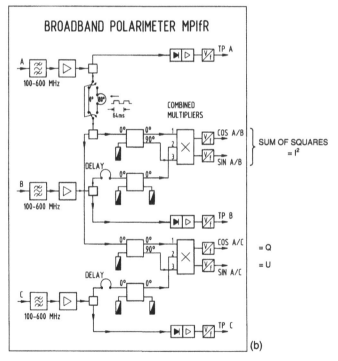

Fig. 6.15 A modern linear polarimeter, with circularly polarized feed, adapted from Turlo *et al.* (1985). (a) Overall receiver system; (b) expansion of the 'polarimeter' module. The TP ('total power') outputs are used to monitor internal signal levels and provide a check on stability of the gains of preceding stages. The Q, U and I^2 outputs are identified; see the footnote to p. 112 for the notation used by the authors.

component must have both the correct polarization and the correct phase difference between the two telescopes for it to lead to maximum output of the correlator). Aperture synthesis and VLBI techniques as such are not the subject of this book. The polarimetry extension may be summarized as follows:

In correlating the outputs from two elementary telescopes of a synthesis or VLBI installation, there is a choice of two orthogonal polarizations in each; therefore four instantaneous combinations can be made, which is just sufficient to determine all four of the Stokes parameter visibilities $\mathscr{I}, \mathscr{Q}, \mathscr{U}, \mathscr{V}$. The time series of $\mathscr{I}, \mathscr{Q}, \mathscr{U}, \mathscr{V}$ visibilities can be Fourier transformed into maps of their corresponding sky distributions I, Q, U, V, after which sky distributions of degree and angle of polarization can be derived.

For a more detailed description, the reader is referred to Christiansen and Högbom (1985, sections 7.13, 7.14) and to Thompson *et al.* (1986, p. 150). The mathematics is all in terms of complex numbers to represent the electric fields and the currents or voltages into which they are translated. This leads to complex correlator outputs, representing 'complex visibilities' containing amplitude and phase of the sinusoid resulting from a source moving through the interferometer antenna pattern. These complex visibilities $\mathscr{I}, \mathscr{Q}, \mathscr{U}, \mathscr{V}$ are Fourier transformed into real sky distributions; the fact that the sky distributions of I, Q, U, V must consist of real numbers translates into hermiticity of the complex visibilities, and this property is used to fill a part of the visibility plane which is not actually observed.

The hardware necessary to measure a complex visibility includes 'fringe-stopping' (which reduces the frequency of the sinusoid to exactly zero for the 'fringe-stopping-centre' of the field and to near zero for other points) and 'complex correlators' consisting of two simple correlators in quadrature, i.e. with 90° phase delay in one of the inputs of one of them.

Morris *et al.* (1964) derive a general expression for the (complex) output r_{mn} of a correlator for the signals from dipoles m and n, with correlator channel gain G_{mn}, the input Stokes visibility vector being $\mathscr{I}, \mathscr{Q}, \mathscr{U}, \mathscr{V}$, with χ and β (fig. 6.13) in each arm adjusted to any value one wishes:

$$
\begin{aligned}
r_{mn} \propto G_{mn} \times \{ &\mathscr{I}[\cos(\chi_m-\chi_n)\cos(\beta_m-\beta_n) + i\sin(\chi_m-\chi_n)\sin(\beta_m+\beta_n)] \\
+ &\mathscr{Q}[\cos(\chi_m+\chi_n)\cos(\beta_m+\beta_n) + i\sin(\chi_m+\chi_n)\sin(\beta_m-\beta_n)] \\
+ &\mathscr{U}[\sin(\chi_m+\chi_n)\cos(\beta_m+\beta_n) - i\cos(\chi_m+\chi_n)\sin(\beta_m-\beta_n)] \\
- &\mathscr{V}[\cos(\chi_m-\chi_n)\sin(\beta_m+\beta_n) + i\sin(\chi_m-\chi_n)\cos(\beta_m-\beta_n)] \}
\end{aligned}
$$

By adjusting the χ and β values, the correlator can be made sensitive to various (linear combinations of) the Stokes parameters. Several examples are discussed in Thompson *et al.* (1986, p. 102 *et seq.*).

A single-dish correlation polarimeter may be regarded as a special case of a two-element polarization interferometer, viz. one with orthogonally polarized channels and zero antenna spacing. This concept is useful when tracing much of the relevant mathematics; note that $\mathscr{I}, \mathscr{Q}, \mathscr{U}, \mathscr{V}$ are now synonymous with I, Q, U, V. Single-dish polarimeters are more sensitive to local interference than are widely spaced correlation interferometers, since the two receiving dipoles are in the same location and spurious local signals are likely to be correlated.

VLBI polarimetry is basically the same as polarimetry with any synthesis instrument, the only difference is in the way the link between the telescopes is implemented. In theory, polarimetry adds a few constraints to the image reconstruction process, such as the 'Stokes criterion' $I^2 \geq Q^2 + U^2 + V^2$. To make use of them, however, would require that the reconstruction takes place in all four Stokes parameters in one combined operation, which may not be worth it for the weak constraints involved; for synthesis polarimetry in general, usable constraints may emerge from full matrix treatment of the entire installation; this approach has only just started (see section 4.4).

Specialized calibration procedures do exist for synthesis polarimetry, but these are highly specific for the instrument and configuration considered. Thompson *et al.* (1986), Fomalont and Perley (1989) and Spoelstra (1992) provide ample detail and references. This is an area in which substantial further development is to be expected, in particular by adapting the matrix methods of chapter 4 to radio correlation interferometry (Sault *et al.* 1995).

With this brief introduction and the previous grounding in polarimetric concepts, the reader should be able to understand original papers in aperture synthesis and VLBI polarimetry. The subject is still very much in development, and interesting new experimental techniques are to be expected, particularly where the extra freedom offered by synthesis methods is integrated within full polarimetry.

6.3 Infrared developments

During the last few years, quality infrared arrays have become available; one result of this has been that CCD polarimetric techniques are being pushed into the infrared. The technique is the same, in principle, as that in the optical region, but the practical components are different and they have to be accommodated within the requirements of an infrared system (e.g. cadmium sulphide wave plates and wire-grid polarizers for 8–13 μm, see Smith *et al.* (1994)).

Infrared instruments generally are entirely contained within a cryostat, which limits the sophistication of mechanical arrangements. A good example is provided by the linear polarimeters of the infrared observatory ISO, for wavelengths between 3 and 240 μm (Klaas *et al.* 1994). These polarimeters are

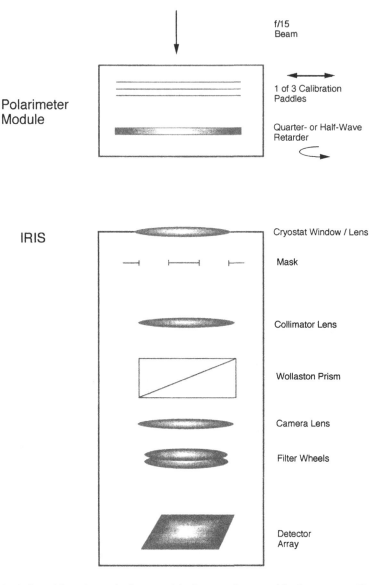

Fig. 6.16 An infrared imaging polarimeter with the retarders outside the cryostat, from Hough *et al.* (1994); IRIS is the Anglo-Australian Telescope's multi-purpose infrared imager/spectrometer. The fact that the (often compound) retarders do not have to be subjected to temperature extremes allows a simpler and more reliable design with less stress-birefringence; having the retarders accessible is also more convenient operationally.

basically of the very simple 'rotating polarizer' type, which relies on calibration rather than modulation for reducing errors. Since the system has about 10% instrumental zeropoint polarization, the final error level is likely to be between 0.1% and 1%.

Hough *et al.* (1994) report on a successful infrared (1–2.4 µm) polarimeter using a warm rotatable halfwave plate outside the cryostat (fig. 6.16). A retarder is not a lossy component, hence its temperature should be irrelevant; to the extent that this is true for a practical optical component (i.e. imaginary part of refractive index negligible for the wavelength range admitted to the detector), this arrangement should work, considerably simplifying construction, hence allowing more sophisticated polarimetric arrangements. It remains to be seen how far into the infrared this technique can be pushed (interestingly, Clemens *et al.* (1990) used it at 1.3 mm).

6.4 (Sub-)millimeter systems

At wavelengths between a few millimetres and a few tenths of a millimetre, instrumental techniques are generally borrowed from both the optical and the radio spectral regions. Polarimetry is no exception; the most common arrangement seems to be a rotating wave plate, followed by a wire-grid polarizing beam-splitter, followed by twin radio receivers. The experimental techniques are developing fast. Relevant papers are Hildebrand *et al.* (1984), Clemens *et al.* (1990), Flett and Murray (1991), Murray *et al.* (1992); for the longer wavelengths, see figure 3 of Baars *et al.* (1994) (the polarizing grids of that figure are used as wavelength-selective – or 'dichroic' – mirror and polarization analyser in one component; for polarimetry a rotatable wave plate can be added).

High accuracy has been achieved in some cases. Clemens *et al.* (1990) quote 0.2% for instrumental polarization at a wavelength of 1.3 mm and estimate their final precision at 0.01 to 0.03%, using the optical technique of a halfwave plate modulating the polarization at about 25 Hz; interestingly, they also mention the 'radio' problem of polarized sidelobes at 0.25 to 1.0%. Component construction depends very much on the wavelength of operation (e.g. a grooved dielectric plate halfwave retarder in Clemens *et al.* (1990)), but the principles of operation are no different from those treated in sections 6.1 and 6.2.

6.5 Ultraviolet systems

To the extent that it exists, astronomical ultraviolet polarimetry is based on optical techniques. From the ozone absorption at about 310 nm downwards, ultraviolet astronomy is space-based, which in practice means that instrumentation has been simple; even the Hubble Space Telescope's (HST) two cameras and High Speed Photometer use fixed analysers, do not have modulators, and the possibility of rotating the instrument is only that of repeat observations at non-standard 'roll-angles' (which are severely restricted because thermal and power management suffer). The Faint Object Spectrograph on HST does have a rotating-waveplate option, and polarimetry accurate to 0.1 or 0.2% has been obtained (Somerville *et al.* (1994); see also section 7.7). With the COSTAR optical correction unit now in use, polarimetry with the HST will suffer from instrumental polarization to some extent; this will have to be calibrated, with a loss of effective observing time. Schmidt *et al.* (1992b) discuss HST standard stars for linear polarimetry from 340 to 880 nm; moderate extrapolation to shorter wavelengths should be possible.

There is no fundamental reason why accurate polarimetry should not be taken down to about 150 nm or even beyond. Polarimetric calibration of a spectrograph for the range 200–450 nm is discussed in Morgan *et al.* (1990), who show that precision methods are feasible. Nordsieck *et al.* (1994a) describe a polarimeter using a rudimentary rotating halfwave modulator over the range 140–320 nm; a precision of order 0.05% is obtained, in spite of tracking problems in the space environment. Given sufficient scientific interest (in this respect, Clayton *et al.* (1992) have shown that there is additional information on the interstellar grains, at least), precise polarimetry down to Lyman-α seems possible: magnesium fluoride has useful transparency down to about 115 nm, and its birefringence is documented at least down to 150 nm (Bennett and Bennett (1978, table 15); similarly for quartz, which stops transmitting at about 150 nm, however); this means that both crystal polarizers and achromatic retarders of reasonable performance can be made, although the bandwidth may be restricted, since the spectral dispersion of material properties is much stronger in the ultraviolet than in the visible.

New developments in ultraviolet polarimetry are reported in a recent conference volume (Fineschi 1994). In one of the papers of that volume, Nordsieck *et al.* (1994b) describe a polarimeter – WISP – for 135 to 260 nm, which uses a large rotatable pneumatically driven stress-birefringent wave plate of calcium fluoride as its modulator; this may be a breakthrough of great significance, since with a change of material and parameters such a component will allow considerable freedom of polarimeter design at other 'optical' wavelengths.

6.6 X-ray systems

X-ray polarimetry as an astrophysical technique is in an early stage of development, due to the severe technological difficulties and the required telescope collecting area. Several attempts at *linear* polarimetry have been made, using sounding rockets and satellites. The only statistically significant detection of polarization is that of the Crab Nebula (fig. 3.6); upper limits for the degree

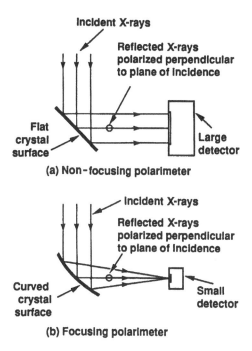

Fig. 6.17 Bragg crystal polarimetry, from Silver *et al.* (1989); rays not reflected pass through the crystal. The 'focusing' configuration (b) is, in fact, telescope and polarizer in one; when a separate telescope provides the focusing, the Bragg reflector can be flat (a). Focusing is important for obtaining the low instrumental background count rate associated with a small detector; lower detector background noise will permit observations of fainter astronomical sources.

of linear polarization exist for several other sources. The existence of a recent technical conference volume on X-ray and ultraviolet polarimetry (Fineschi 1994) is sufficient evidence of continuing interest and development; in particular, the development of X-ray retarders may make *circular* polarimetry a possibility in due course.

Existing X-ray polarimeters are basically of a simple 'optical' type, consisting of a linear polarizer followed by or combined with detectors, the whole assembly (with or without the telescope) rotated about the optical axis. The polarimeters that can be used at the focus of a telescope (soft X-rays, energies up to $\approx 15\,\text{keV}$) are of two designs, each used for a different purpose:

- Bragg crystal polarimeter. In such an instrument (fig. 6.17), Bragg reflection is used (constructive interference of the waves scattered from successive atomic layers within the crystal). For a given angle of incidence, a very narrow range of wavelengths is reflected. A graphite Bragg reflector is usually made up of many micro-crystals with orientations differing by a

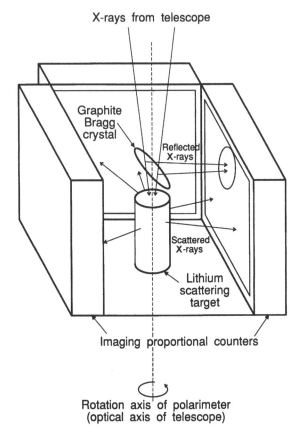

Fig. 6.18 The Stellar X-ray Polarimeter, SXRP (there are, in fact, four detector panels, the front panel is not shown). Bragg (graphite) and Thomson (lithium) scattering polarimeters combined into a single instrument. The graphite crystal behaves as a single-beam narrow-band reflection linear polarizer for two wavelengths; radiation of other wavelengths passes straight through the graphite and is used by the lithium polarimeter. The lithium polarimeter is of the double-beam type: the polarized component appears alternately in one or the other pair of detectors as the instrument is rotated at ≈ 1 rev/min. The lithium polarimeter can achieve moderate spectral resolution for strong signals, by energy-binning the output pulses of the proportional counter. Figure kindly provided by P. Kaaret.

few degrees; each micro-crystal then reflects a slightly different wavelength and the resultant bandwidth has a more useful value (without the loss of effective area that an array of larger mono-crystals would produce: radiation passes straight through the matrix until it finds a micro-crystal of the correct orientation, thus the entire reflector area contributes at all wavelengths within the passband determined by the distribution of orientations of the micro-crystals). The first-order band for graphite is at 2.6 keV, and the OSO-8 Crab Nebula observation in fig. 3.6 was made with such a polarimeter operating at 15% bandwidths centred on 2.6 and 5.2 keV.

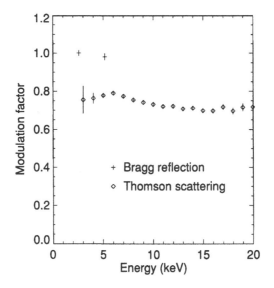

Fig. 6.19 Polarimetric efficiency of the lithium Thomson scattering (◇) and graphite Bragg (+) polarimeters for SXRP; the useful energy range of the system will be from 2.6 to about 15 keV, the telescope limiting the performance at the high-energy end. Figure kindly provided by P. Kaaret and R. Elsner; data from Elsner *et al.* (1990).

- Thomson scattering polarimeter. When a material scatters X-rays efficiently (compared with absorbing them), this can be used to construct broad-band polarimeters, since linearly polarized radiation is scattered preferentially at right angles to the direction of vibration of the electric vector (see fig. 6.18; further development of such techniques for hard X-rays (50–500 keV), using fibre scintillators, is proposed by Costa *et al.* (1994)).

Fig. 6.18 shows the combined Bragg and Thomson polarimeters of SXRP, which, associated with the SODART telescope (150 nested stretched-foil grazing-incidence mirrors) will allow sensitive linear X-ray polarimetry on the Spectrum-X-Gamma mission. Fig. 6.19 shows the polarimetric efficiency of these instruments. A minimum detectable degree of polarization of about 0.5% is expected for 30 hours of integration on the Crab Nebula (Kaaret *et al.* 1994); clearly X-ray polarimetry will be largely photon-limited for some time to come.

A completely new type of *imaging* linear *spectro*-polarimeter for X-rays of a few tens of kilo-electron-volts may result from the work reported by Tsunemi *et al.* (1994). Charge clouds from polarized X-ray photons striking a CCD in an imaging setup are found to be elongated along the electric field vector of the incident radiation, while the number of electrons in each cloud depends on the energy of the X-rays, as usual. The first practical test of this principle was effective from about 20 to 40 keV, with polarimetric efficiency of order 0.1 or 0.2; the technique seems capable of considerable improvement. A test of a somewhat similar idea is reported by Austin *et al.* (1994) for the 40–100 keV range: the ejection directions of K-shell photo-electrons in a proportional ionization chamber are correlated with the linear polarization of the incident X-rays. Via electron multiplication leading to optical emission, followed by CCD readout and software analysis of the optical image, the original ejection direction is determined. A measured polarimetric efficiency of 30% at 60 keV is reported for the laboratory test. It will be some years yet before these new ideas are translated into practical astronomical polarimeters.

6.7 γ-ray systems

Polarimetry at γ-ray wavelengths has not yet started. Kotov (1988) reports acceptable modulation efficiencies between 25 and 90% for various interactions with γ-ray detectors that could be used to measure γ-ray (linear) polarization and explores the possibility of using the COMPTEL telescope on the γ-ray observatory GRO. Kotov concludes that marginal observations (3σ) at degrees of polarization of 15 to 50% will take integration times of one to three months! Enough said for the moment; astronomical urgency will have to be shown first.

7

Case studies

In this final chapter, several original papers from the literature are introduced, primarily as illustrations of modern instrumentation. The focus of this book being the *measurement* of polarization, the astronomy involved was a secondary consideration in selecting these particular papers from the wide range available. Readers are urged to test their grasp of polarimetric fundamentals by selecting a dozen or so further papers from those listed in the subject index of the more recent volumes of *Astronomy and Astrophysics Abstracts* under 'polarimeters', 'polarimetry' or 'polarization'.

7.1 Multi-channel optical polarimetry using photomultipliers

A suitable example of optical polarimetry by the 'classical' technique of 100 Hz modulation and photomultiplier detectors is given in Können and Tinbergen (1991) and Können *et al.* (1993). It concerns an attempt to detect ice crystals in the upper parts of the Venus atmosphere by using the polarization peak at the 22° halo angle as a diagnostic. A large body of earlier Venus polarimetry exists, and scientific results derived from it are reviewed in Van de Hulst (1980, section 18.1.5 and references therein).

The terrestrial 22° halo and related phenomena owe their polarization to birefringence of the ice crystals that produce the halo. These crystals operate as 60° prisms, deviating the light from the Sun by an amount depending on the refractive index of the ice, hence by an amount which depends on the polarization of the light. The Sun's image has a sufficiently sharp edge for noticeable separation of the haloes as seen in two orthogonal linear polarizations (parallel and perpendicular to the scattering plane). The theory of this is well established; Können and Tinbergen (1991) document measurements on such a halo and a mock Sun. These measurements fit the theory to perfection.

When the Sun–Venus–Earth angle passes through 22°, one would expect

any Venus ice crystals to cause a short-lived reduction of the otherwise slowly
varying Venus polarization (which is due to scattering at the sulphuric acid
droplets in the clouds); the timing of such reduction will depend on the
wavelength in a predictable manner.

This basic idea was put to the test using a multi-channel photomultiplier polarimeter with 100 Hz modulation; technical details can be found in Tinbergen
(1987b). The astronomically relevant part of the investigation is reported in
Können *et al.* (1993). When the halo angle is passed, Venus is only about 15°
from the Sun, so daytime observations are necessary. The precautions taken
to eliminate errors by scattered light and background polarization (bright blue
sky) are discussed, as are the statistical errors obtained at several levels of
combination of the individual observations. The reduction procedure is summarized below; it is typical of stellar/planetary polarimetry using single-pixel
detectors (only the last two steps are specific to the scientific project).

- Subtract background polarized component (in I, Q, U, in the instrument
  coordinate system) via interpolation of the 'sky' observations (which include
  light scattered via the floor etc. and then via the white inside of the dome).
- Form $Q/I, U/I$.
- Subtract instrumental polarization (determined by night-time observations
  of zero-polarization stars).
- Correct for wavelength dependence of polarization angle, caused by superachromatic halfwave plate modulator.
- Transform to coordinate system defined by the Sun–Venus–Earth plane
  (astronomical system).
- Plot $Q/I, U/I$ against scattering angle. U/I should be nearly zero and show
  no particular structure near 22°.

Interpretation involves the dependence, on scattering angle and wavelength,
of Q/I (astronomical coordinate system). As temporal changes in the Venus
atmosphere contaminate the expected variation of polarization with scattering
angle, this interpretation is complicated and is not really relevant to this book;
for details, the reader is referred to Können *et al.* (1993).

7.2 Optical spectro-polarimetry using CCD detectors

Early examples of CCD spectro-polarimetry using the *common-user* spectrograph shown in fig. 6.5 are those reported by Rutten and Dhillon (1992) and
by Schild and Schmid (1992) (see fig. 3.9). This well-documented spectrograph
is a red/blue/faint-object triple instrument using a common-slit facility which
includes the polarization module; it is described in detail in its manual (Clegg

et al. 1992), while polarimetric application is the subject of a separate manual (Tinbergen and Rutten 1992). The polarimetric regime is used routinely, for both linear and circular polarimetry in spectral and in imaging mode; long-slit spectral mode is a third option.

CCD spectro-polarimetry in many ways is just multi-channel polarimetry, as discussed in the previous section. The differences are in the modulation rate (very slow), double-beam operation and in the data reduction (which must cope with roughly 1000 wavelength channels). A complication of the William Herschel Telescope (a complication which has its uses in polarimetry) is that it has an alt-azimuth mounting, with an instrument rotator to allow one to keep the instrument axes aligned with the equatorial coordinate system. In polarimetry, one generally prefers not to disturb the relationship between telescope and instrument, as that would change the instrumental polarization zeropoint. The compromise one uses in actual practice is discussed by Tinbergen and Rutten (1992, p. 18); other sections of that instrument manual worth looking at are those on scattered light and on photon shot noise in polarimetry.

7.3 Solar imaging spectro-polarimetry by advanced CCD methods

By far the most advanced polarimetry being attempted in the optical wavelength region is that planned for the Large Earth-based Solar Telescope LEST (fig. 5.1). An introduction to this planned 2.4 m helium-filled tower telescope (for optimal atmospheric seeing) is given by Engvold (1992). Polarimetry is an essential option of almost all of its instrumentation. For the polarimetric aims, I quote from the LEST Foundation Annual Report 1993, section 3:

The scientific aims of LEST require polarization measurements with high sensitivity and accuracy...polarimetry system needs to record all four Stokes parameters over the full wavelength range that the core telescope provides (0.3–2.7 µm)...polarimetry is an integral part of LEST...in particular the near-infrared, where the ratio between Zeeman splitting and Doppler width increases proportional to wavelength...observations of intra-network fields in the photosphere and weak chromospheric fields require high polarimetric sensitivity...determinations of the field orientation require high accuracy of, in particular, the linear polarization...most specifications must be met by all systems: • sensitivity determined by photon statistics, at least 10^{-4} of the total intensity • accuracy (cross-talk between Stokes components, off-diagonal elements in the Mueller matrix of the instrument) better than 10^{-3} • wavelength coverage 0.3 to 2.7 µm; simultaneous wavelength coverage over at least 1000 Å • photon flux not reduced by more than a factor of 3 by the polarimeter • up to 10 full frames per second...the main problem with the LEST polarimetry system is due to the core telescope itself. Birefringence due to remaining stress in the front window of LEST is a major source of cross-talk between the Stokes param-

128 *Case studies*

eters (at least for the visible). Much of the engineering work...should be devoted to the window behaviour and the polarization calibration of the telescope.

The polarization behaviour of the telescope itself is discussed by McGuire and Chipman (1988); Keller *et al.* (1992) is a design review of a prototype imaging full-Stokes polarimeter for visible wavelengths; Povel *et al.* (1989) and Stenflo *et al.* (1992) report on the development of some important details of such a polarimeter. Together, these papers take the reader right up to the state-of-the-art in optical polarimetry.

7.4 Single-dish radio-polarimetry; atmospheric compensation

The 2.8 cm study of M51 by Neininger (1992) is a very good example of modern radio-polarimetry (see fig. 3.4). The sophisticated receiving system which made the study possible is described briefly in Schmidt *et al.* (1993); a similar 9 mm system is described in more detail by Morsi and Reich (1986).

The polarization of M51 has been studied, at many different wavelengths, to determine the magnetic field structure and to put constraints on dynamo models for the origin of the large-scale magnetic field. Interpretation of optical polarimetry of external galaxies is always confused by scattered radiation, at radio wavelengths the problem is Faraday rotation. This study at 2.8 cm manages to avoid both problems, Faraday rotation being less (probably considerably less) than 5° throughout the field. The multi-beam receiving system is sensitive enough for this application and the beam-switching eliminates most of the atmospheric noise.

Although Neininger used a technologically advanced observing system, most of his paper is devoted to a discussion of the results and their implications for models of the origin of the magnetic field. This is a tribute both to the quality of the receivers and to the contribution of polarimetry to astrophysics.

7.5 Radio synthesis array imaging polarimetry

The study by Wieringa *et al.* (1993) is a suitable vehicle for examining the strengths and limitations of synthesis polarimetry. The reader can gain a good impression of the importance and subtlety of data processing in a mature synthesis system, and of the extra freedom offered by a synthesis telescope for designing calibration procedures (see Sault *et al.* (1995) for more recent developments).

Fig. 7.1 shows a WSRT field on which sensitive galactic-foreground continuum polarimetry was performed (partly serendipitously, while pursuing a cosmological goal, viz. attempted detection of 21 cm line radiation at a redshift

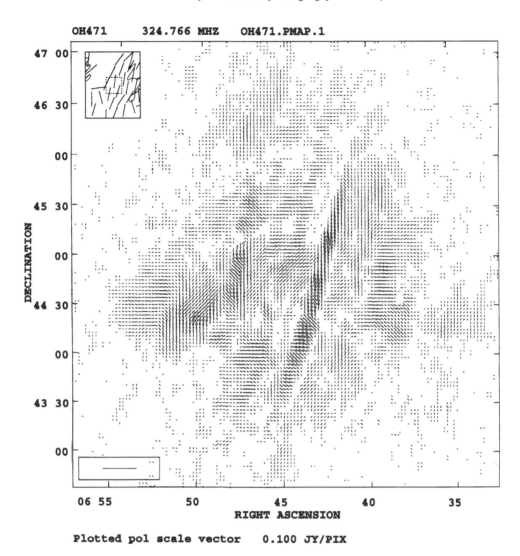

Fig. 7.1 Synthesis array polarimetry, at 325 MHz, of local galactic continuum radiation (Wieringa *et al.* 1993); plane of vibration of the *electric* vector, small-scale structure only, strongest point sources removed. The scale of the polarization lines is indicated by the boxed line at lower left (0.1 Jy/pixel). The direction of the large-scale *magnetic field*, as deduced from multi-frequency single-dish observations (Spoelstra 1984), is shown in the inset top left, with approximate size and location of the synthesis frame sketched in (dotted square).

of 3.3). In studying this material, one should bear in mind that the smallest spacings of the interferometers were absent, so that (polarization) structure of larger scale than about 1° has been filtered out. However, this large-scale structure is known approximately from the multi-frequency single-dish study by Spoelstra (1984); no observations at 325 MHz exist, but an extrapolation can be made, see the maps of polarization angle and polarization brightness temperature (figures 7 and 8 of Brouw and Spoelstra 1976); the large-scale magnetic field direction (at right angles to the vibration direction of the electric field for 'as-emitted' synchrotron radiation) is indicated in fig. 7.1.

Wieringa *et al.* (1993) note that the *I* map does not contain features corresponding to the polarization structure, therefore the small-scale structure in the polarization must be caused by small-scale structure in polarization conversion (without loss of *I*, i.e. non-zero off-diagonal elements only in rows or columns 2, 3 and 4 of the Mueller matrix of the intervening medium). Highly structured foreground Faraday rotation of radiation from a spatially very smooth polarized background source is the suggested mechanism. This background polarized component with only large-scale structure could be identical to that documented by Spoelstra.

7.6 Sub-millimetre polarimetry

The hybrid nature of (sub-)millimetre polarimetry is well illustrated by Clemens *et al.* (1990). In their instrument, a rotating halfwave modulator at room temperature is combined with a cooled wire-grid polarization beam-splitter (dual-beam analyser) and an existing cooled 1.3 mm dual radio receiver. A level of accuracy in degree of polarization of about 0.02% was reached, which is heroic considering that 'even for a relatively bright source like the Orion cloud core, the unpolarized flux represents only 0.05% of the receiver plus sky noise...this drives the dynamic range to almost 10^6...the cost of the project had to be essentially zero'. The paper discusses typically optical items like halfwave plate (grooved Plexiglass) modulation and equally typical radio concerns like sidelobe polarization and linearity of the electronics, while typically far-infrared preoccupation with sky opacity and background is never far out of the picture, either. The superb experimental skill involved is evident from the overview table 1.3, in which the 1.3 mm entry refers to Clemens *et al.* (1990).

7.7 Ultraviolet polarimetry

Very little ultraviolet polarimetry exists at present (see section 6.5), but the HST will help to put this right. One should beware of instrumental polarization effects in COSTAR-assisted data (however, pre-COSTAR polarimetry also has its problems; see Somerville *et al.* (1994, section 3)).

The interstellar polarization curve for the star HD161056 is now documented over a 25:1 range in wavelength , with really astonishing accuracy (fig. 3.11). The relevant papers are listed in the references of Somerville *et al.* (1994). Although the HST/FOS polarimetry does not seem to be of quite the same quality as that of WUPPE (Clayton *et al.* 1992, Nordsieck *et al.* 1994a), the very complete documentation of the interstellar polarization of the light from one particular star is worth studying. The main concern of ultraviolet polarimetry in this field of astronomy is the search for deviations from the empirical 'Serkowski' curve and for more stringent constraints on grain models and size distributions.

Exercises

These exercises are intended both for private study and for classroom discussions; they raise questions and topics that keep on surfacing in the author's own mind. Many of the exercises are rather open-ended, so only 'hints' are provided (as opposed to 'answers to problems').

Chapter 1

1.1 Look up 'polarimetry' and/or 'polarization' in the latest volume of *Astronomy and Astrophysics Abstracts*. Take any astronomical topic you fancy and, using the references in the *Abstracts* and in the papers you find, trace the use of polarization in your chosen topic back to its origin. How does the original paper you find fit into the 'milestones'? Take several other astronomical topics and repeat the exercise; can you improve on tables 1.2 and 1.3? Write summaries of the papers for future reference; you will enjoy reading them later.

1.2 Take two Polaroids and look at a bright source of light through the two in series. Now rotate one of them. Do you get complete extinction at some angle of rotation? If not, try to explain what you do see.

Now discard one of the Polaroids and look through the other at the world around you, rotating the Polaroid as you do so. Examine, in particular, blue sky about 90° from the Sun, clear sky close to the Full Moon and at about 90° from it, a rainbow, a mock Sun, the 22° halo, light reflected off a puddle in the road and off a nice shiny car. Make notes on what you observe.

1.3 Take two Polaroids, rotate one for (near-)extinction. Now insert a third Polaroid and/or any other transparent material into the space between the Polaroids. Rotate the middle component. Try to explain qualitatively what you see.

Now use a piece of plastic foil as the middle component and *stretch* it. Also use a piece of Perspex and *bend* it. Note down everything you observe, for future reference (after chapter 6 you should be able to analyse what you have seen).

Chapter 2

2.1 Can LHC polarization change 'very slowly' into RHC? If so, how? If not, why not?

2.2 Which word *should* Sir George Gabriel Stokes (or his CUP support staff) have used instead of 'oppositely' (marked sic! in the quotation on p. 17 from Stokes (1901))?

2.3 Can monochromatic radiation be unpolarized? Discuss the reasons for whatever answer you give.

2.4 How would you describe the relationship between two polarization states represented by opposite ends of any diameter of the Poincaré sphere?

2.5 How would you describe the polarization state represented by a point within the Poincaré sphere that describes a circle in the equatorial plane, at a uniform speed of 1 rotation per second and at a radius of 0.5? Answer the same question for a circle plane at right angles to the equatorial plane and for a circle plane at an arbitrary angle to the equatorial plane.

2.6 In the production of *un*polarized radiation, 'all values of β and χ will occur'. What does this imply for the polarization ellipse of the radiation, and can you determine from that geometrical picture that the time average must indeed be 'zero polarization'?

2.7 'If polarization measurements yield a null result no matter what polarization form one tries to detect, the radiation under scrutiny is said to be unpolarized.' What is the minimum number of polarization measurements required to determine that a particular beam of radiation is unpolarized? Assume that 'very slow variations' do not occur, i.e. that the statistical properties of the signal do not evolve.

2.8 During the author's initiation as a polarimetrist (Westerhout *et al.* 1962), a sequence of 12 hours of observations of the celestial North Pole yielded results schematically represented in fig. Ex.1. What do you think should be the interpretation of the ellipse with axial ratio close to 2 and major axis aligned more or less radially?

2.9 In quasi-monochromatic 100% linearly polarized radiation, the orientation of the plane of vibration may change *very slowly*. One may also conceive of quasi-monochromatic linearly polarized radiation for which

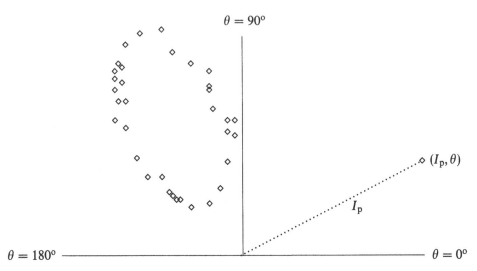

Fig. Ex.1 The author's first steps as a polarimetrist. This figure shows, plotted in polar coordinates, 12 hours of observations at the celestial (equatorial) North Pole, with a stationary alt-azimuth telescope. I_p is the polarized intensity and θ is the polarization angle, as deduced from the sinusoidal variation of the 'polarization' output of the receiver (see fig. 6.12).

phase and amplitude change only *very slowly*, while, on the other hand, the *orientation* changes *slowly* in a random manner. How would you describe the state of polarization of such radiation? Would you say the concept of *slowly* changing orientation is very useful?

Chapter 3

3.1 Why would you, in general, expect observed polarization to rise as the resolution of an observation is increased (distinguish between spectral and spatial resolution)? Can you think of exceptions to this general rule?

3.2 Having observed that the light from a sunspot umbra is, in general, polarized at some (low) level, how would you attempt to distinguish between scattering and magnetic fields as the origin of this?

3.3 Sunspots often occur in pairs, of opposite magnetic polarity. What is the relationship between the polarization of normal Zeeman triplets from the members of such a pair? Would you expect any observable effect (in polarization and/or in some other way) for a remote (i.e. point source) star with many such pairs of spots on its surface?

3.4 Red supergiant stars such as Betelgeuse and Antares show variable, wave-length-dependent, linear polarization which is attributed to asymmetries in the scattering of photospheric light by the tenuous outer layers of the atmosphere. The asymmetry could be in the photospheric light distribution or in the distribution of scatterers. Convince yourself that either of these possibilities will indeed generate linear polarization; how would you attempt to distinguish between them?

3.5 If the charge-to-mass ratio of the electron had been much larger than it actually is, the dispersion with frequency of the pulse arrival times and of the Faraday rotations would also have been (very) much larger. How would this have affected the discovery of (the polarization of) radio pulsars? Given the theoretical concept of neutron stars and the oblique-rotator model for magnetic A stars, polarization of pulsed radio emission might nevertheless have been predicted. How would you set out to verify such a prediction?

3.6 In the universe of the previous exercise, would you expect linear polarization of extragalactic radio sources to have been discovered at all? If so, how and when? If not, why not?

3.7 Polarization of radiation always implies some sort of asymmetry somewhere. Is it possible that a *velocity* of the source relative to the observer could generate or modify polarization (transverse velocity $\Longrightarrow$ linear polarization, longitudinal $\Longrightarrow$ circular)? In particular, can radiation that is emitted unpolarized in the frame of the source appear polarized to an observer, by the mere fact of relativistic motion of source relative to observer?

Chapter 4

4.1 In example 3 of section 4.1, a Mueller matrix is derived for a linear polarizer in general orientation θ. Does it work out all right for the $-Q$ polarizer of example 2? And what is the Mueller matrix for $\theta = 180°$, or $\theta = \pm 45°$? Verify that your matrices actually do what you would expect, for several relevant input Stokes vectors (such as: linear polarization in several orientations, unpolarized and circularly polarized radiation, partially and elliptically polarized radiation).

4.2 What is the Mueller matrix of a halfwave plate in principal orientation? What does such a halfwave plate do to circular polarization? And what does it do to linear polarization at an angle θ?

4.3 What is the Mueller matrix of a halfwave plate at an arbitrary orientation ψ? What is the output for constant input linear polarization as the halfwave plate is rotated at a uniform rate? And what is the output in that case for constant input circular polarization?

4.4 Prove to yourself that a quarterwave plate oriented at an angle θ has the following Mueller matrix:

$$
\begin{pmatrix}
1 & 0 & 0 & 0 \\
0 & (\cos 2\theta)^2 & \cos 2\theta \cdot \sin 2\theta & -\sin 2\theta \\
0 & \cos 2\theta \cdot \sin 2\theta & (\sin 2\theta)^2 & \cos 2\theta \\
0 & \sin 2\theta & -\cos 2\theta & 0
\end{pmatrix}
$$

What does such a quarterwave plate do to Q when $\theta = \pm 45°$? And to V?

4.5 The usual way to produce circularly polarized light is to use a linear polarizer, followed by a quarterwave plate at an orientation of 45°; the combination is an 'inhomogeneous circular polarizer'. The combination has practical disadvantages, since the quarterwave plate is never completely achromatic. Fresnel constructed a two-beam *homogeneous* circular polarizer by using alternate prisms of right-handed and left-handed quartz crystals (Hecht and Zajac 1974, figure 8.59; Collett 1993, figure A-3). Write down the Mueller matrix for each of these two types of circular polarizer and discuss the differences. Stenflo (1994, p. 38) introduces a 'circular polarization filter' consisting of two quarterwave plates with parallel orientation, with a linear polarizer at 45° sandwiched between them. Use Mueller matrices to discuss the behaviour of this component in relation to that of the previous two.

4.6 Given that Faraday rotation is proportional to the square of the wavelength, what is the Mueller matrix at wavelength λ metres for a medium which rotates the plane of polarization by x radians at a wavelength of μ metres? How does this medium affect circular polarization?

4.7 In the (rarely used) complex $Q + iU$ plane mentioned in section 2.6, how would you represent the integrated linearly polarized radiation received from the volume within the telescope beam, assuming linearly polarized emission and a thermal plasma causing Faraday rotation, both distributed along the line of sight? Does the form of this integral remind you of anything? Can you use that insight in any practical way?

4.8 In the (more usual) complex plane of section 4.6, states of polarization are characterized by only *two* parameters. How is this related to the statement that the *four* Stokes parameters provide a complete description of polarized radiation?

4.9 Collett (1993, p. 75) gives the Mueller matrix of a linear retarder in principal orientation as

$$\begin{pmatrix} 1 & 0 & 0 & 0 \\ 0 & 1 & 0 & 0 \\ 0 & 0 & \cos\phi & -\sin\phi \\ 0 & 0 & \sin\phi & \cos\phi \end{pmatrix}$$

Compare this with our example 4 in section 4.1. Why the sign differences?

4.10 If you have access to computer facilities with matrix manipulation and plotting routines, try the following design exercise (which took the author weeks in 1972, writing his own matrix multiplication subroutines and plotting the results by hand): Pancharatnam's (1955b) recipe for an achromatic halfwave (linear) retarder is to use a stack of three identical simple (linear) halfwave plates, the outer two having their 'fast' axes parallel, while the central one has its fast axis rotated through an angle ψ (a little less than 60°) with respect to the outer two. By matrix multiplication and by producing as a display a 4 × 4 array of spectra for the matrix of the complete stack (as in fig. 4.1), investigate what happens when you let ψ take the values $52°, 56°, 58°, 59°$ and $59°5$. Note that the abscissa in the display of fig. 4.1 is *relative retardation* (a monotonically and non-linearly decreasing function of wavelength), with the value 1 referring to a wavelength somewhat blueward of the centre of the design range; refer to Tinbergen (1973) if you need more detail.

If you have managed this exercise, you might like to investigate manufacturing tolerances on the thickness of (a) the central plate and (b) one of the outer plates; this will give you some feeling for the kind of work an optical instrument designer spends most of his time on.

4.11 Pancharatnam (1955b) states that an assembly of three linear retarders, the outer two being quarterwave plates with their axes parallel and the central retarder a halfwave plate, can act as a linear retarder with its axes at 45° to those of the outer pair; the effective retardation of the combination is four times the angle which the axes of the central plate make with respect to those of the outer two. Prove this for yourself using Mueller matrices. Can you suggest practical applications of this?

Chapter 5

5.1 The new 'multi-frequency front-ends' of the component dishes of the Westerbork Radio Synthesis Telescope can rotate the entire focus assembly to position the front-end unit for the chosen observing frequency on to the optical axis of the mirror. This is more complicated than permanently mounting the several front-end units side by side in the focal plane and it does not allow *simultaneous* multi-frequency observations (except at widely separated frequencies, for which two or three units can be concentric with each other). What do you think could be one reason for this design choice? What does your answer imply for the Arecibo telescope, where the mirror is stationary and the focus box is moved to follow a source as the sky sweeps past?

5.2 Having redrawn fig. Ex.1 in exercise 2.8, address the main question of how to interpret the ellipse (now a circle).

5.3 The spectro-polarimeter of fig. 6.5 uses gratings of the type illustrated in fig. 5.7. In view of the inherently erratic polarization behaviour of such gratings, can one trust the polarization spectra obtained? Indicate when your answer may break down.

5.4 'A spectrograph on a Nasmyth platform will allow reliable Zeeman polarimetry, provided the right precautions are taken.' Using the fact that an oblique mirror acts simultaneously as a partial linear polarizer and a linear retarder (the axes for both actions being aligned and both actions being slow functions of wavelength), consider the polarimetric errors introduced by the Nasmyth mirror. Devise ways to minimize the effect of such errors on the Zeeman measurements, using Mueller matrix treatment if possible. Identify 'the right precautions'.

5.5 It has been claimed that, by determining all four Stokes parameters for each of four calibration sources in the sky, one may determine all 16 elements of the Mueller matrix of the telescope. One can then invert this matrix and compute the true polarization of any other source for which one has obtained the measured values of all four Stokes parameters. What are the implicit assumptions in this statement and are they likely to be realized? Can you indicate restrictions on the calibration sources? Distinguish between the following types of telescope: (a) optical, (b) radio: single-dish, and (c) radio: synthesis.

5.6 To obtain a low level of (grating) sidelobes with a synthesis radio telescope, one commonly repeats observations of a given field with a different set of spacings between the individual dishes. Given the variability of the Earth's ionosphere, what do you think the influence will be on polarimetry carried out in this way, and how would you attempt to minimize such influence?

Chapter 6

6.1 In the sine-wave modulator of fig. 6.1, how many periods does the sine wave go through for one revolution of the halfwave plate?

6.2 The sine-wave and square-wave modulators of fig. 6.1 can both be used for linear or circular polarization, by retaining or removing component 1. Discuss the modulation efficiency (and the consequent effective efficiency of telescope use) of these alternative modulators, for the following cases:
(a) circular polarization;
(b) linear polarization of unknown polarization angle;
(c) linear polarization of known position angle.

6.3 Prove to yourself that, in the receiver system of fig. 6.12, the switched halfwave retarder is not needed for the basic action of the circuits. Why do you think the designer has added the complication of synchronized modulation and demodulation?

6.4 What is the function of the signal-frequency noise source in fig. 6.14? And of the circulators?

6.5 Single-dish radio telescopes receive a significant fraction of their signal from directions outside the main beam. To correct a raw sky map for such radiation, one needs an approximation to this sky map and a measurement of the antenna pattern; these data can be obtained nowadays, and in principle the correction can be carried out. To apply such a method properly to 21 cm Zeeman emission measurements, one would need a full Mueller matrix antenna pattern and a full Stokes vector sky map, both as a function of radial velocity (frequency). This sounds like a stupendous problem; by considering the relative size of the various combinations of Stokes parameters and matrix elements, can you reduce the problem to manageable proportions?

6.6 Synthesis radio telescopes are less susceptible than single-dish instruments to spurious 'polarized' signals caused by radiation entering through the sidelobes. What is the reason for this relative immunity and can you think of exceptions?

6.7 Suppose that, in the universe of exercises 3.5 and 3.6, the Crab pulsar had been successfully detected at X-ray and optical wavelengths. Given the extreme sensitivity of Faraday rotation to longitudinal magnetic fields, attempts to measure galactic Faraday rotations would be undertaken at some point and (radio polarization of) pulsars would have been proposed as a consequence of the oblique-rotator model for the Crab pulsar. Specify design aims for a radio system (telescope, receivers, computers) to (a) confirm the existence and measure the polarization of decimetric radio emission from the Crab pulsar, and (b) search at decimetric wavelengths for other examples of the pulsar phenomenon and, if such are found, to measure their polarization. You may assume that sufficient computer power can be made available for your reduction and instrument-control needs.

Chapter 7

7.1 Scan the index of this book and look up any item that still looks unfamiliar.

7.2 Read again the most recent paper on each of the topics you selected for exercise 1.1. You should be able to follow all the polarimetric aspects now, in detail. Compare your understanding now with your previous notes.

Hints for exercises

Chapter 2

2.1 Fully polarized, unpolarized?

2.2 Definition of monochromatic?

2.3 β?

2.7 $Q, U, V = 0, \quad I \neq 0.$

2.8 This way of plotting is not necessarily wrong, but it ignores the true-vector interpretation of the Q, U number-pair. A similar case found in the literature included a vector (with arrowhead) from the origin to the centre of the ellipse. That *is* wrong; why? Similarly, what is wrong with calling fig. 3.7 a 'vector map'? Redraw fig. Ex.1 in a more useful representation.

2.9 Averaging in definition of Stokes parameters. Degree of polarization.

Chapter 3

3.1 Depolarization = ??? of polarization.

3.2 Geometric configuration, type of polarization, spectral dependence.

3.3 (Inverse) Zeeman effect/Jenkins and White (1950); rotation; Doppler shift.

3.5 Spectral region, bandwidth, dispersion compensation, Lang (1980).

3.6 Faraday rotation in the Earth's ionosphere, the Galaxy and in the source itself; technical developments.

3.7 This problem illustrates how a hunch based on symmetry needs the support of a detailed mechanism. I am assured, by those more versed in relativity than I am, that the degree of circular polarization and the degree of linear polarization are separately invariant under Lorentz transformations (Landau and Lifshitz 1975, p. 123). Within the invariant degree of linear polarization, the polarization angle will, in general, change from one frame to another (as will the perceived direction of the beam, the flux density and the wavelength; these, however, have nothing to do with polarization as such). But unpolarized radiation will be unpolarized to all observers.

Chapter 4

4.3   It's all in the matrix, y'know!

4.4   Now compare with fig. 4.1.

4.5   Fig. 4.1 and table 4.2; also Shurcliff (1962).

4.6   Eigenmodes?

4.7 $F_\lambda = \lambda^2 \xi$. $W(\lambda^2) = \int_{-\infty}^{+\infty} w(\xi) e^{2i\lambda^2 \xi}\, d\xi$.   Can you define in words what $W(\lambda^2) = Q + iU$ and $w(\xi)$ stand for ? Total volume within antenna beam. See Burn (1966, p. 71).

4.8   Degree of polarization. Flow of radiant energy. Polarization state.

4.9   Compare Collett (1993) with Clarke and Grainger (1971, p 37).

4.11   Axes of combination at 45° to those of the outer pair. Would an achromatic but variable retarder be useful? If so, how would you construct one?

Chapter 5

5.2   Section 5.5.4.

5.3   Modulation, retarder, stability, signal strength.

5.5   Instrumental stability, conditions on matrix and polarization of the calibration sources.

5.6   What information does one need to correct the effects and how does one get it (from own instruments, from those of others) ?

Chapter 6

6.4   Correlated noise.

6.5   Zeeman measurements use polarization of spectral lines. Spectral structure of polarized sidelobes? Polarization of 21 cm line, averaged over sky? Strong point sources?

6.6   Correlated signals.

References

Aitken, D.K., Roche, P.R., Bailey, J.A., Briggs, G.P., Hough, J.H. and Thomas, J.A. (1986). Infrared spectropolarimetry of the galactic centre: magnetic alignment in the discrete sources, *Mon. Not. Roy. Astron. Soc.* **218**, 363–84.

Angel, R. (1978). Magnetic white dwarfs, *Ann. Rev. Astron. Astrophys.* **16**, 487–519.

Antonucci, R.R.J. and Miller, J.S. (1985). Spectropolarimetry and the nature of NGC 1068, *Astrophys. J.* **297**, 621–32.

Aspin, C., Rayner, J.T., McLean, I.S. and Hayashi, S.S. (1990). Infrared imaging polarimetry and photometry of S106, *Mon. Not. Roy. Astron. Soc.* **246**, 565–75.

Austin, R.A., Minamitani, T. and Ramsey, B.D. (1994). Development of a hard X-ray imaging polarimeter, in *X-Ray and Ultraviolet Polarimetry*, ed. S. Fineschi, Proc. SPIE **2010**, pp. 118–25 (SPIE, Bellingham).

Baars, J.W.M., Greve, A., Hein, H., Morris, D., Penalver, J. and Thum, C. (1994). Design parameters and measured performance of the IRAM 30-m millimeter radio telescope, *Proc. IEEE* **82**, 687–95.

Bailey, J.A. (1989). *TSP – A time series/polarimetry package* (Starlink User Note 66.1, Rutherford Appleton Laboratory, Didcot, UK).

Beck, R. (1993). Galactic dynamos – a challenge for observers, in *The Cosmic Dynamo*, eds. F. Krause, K.-H. Rädler and G. Rüdiger, IAU Symp. **157** (Kluwer Academic Publishers, Dordrecht).

Bennett, J.M. and Bennett, H.E. (1978). Polarization, in *Handbook of Optics*, eds. W.G. Driscoll and W. Vaughan (McGraw-Hill, New York).

Berkhuijsen, E.M. (1975). A consistent scheme of definitions of polarisation brightness temperature and brightness temperature, *Astron. Astrophys.* **40**, 311–16.

Berkhuijsen, E.M., Brouw, W.N., Muller, C.A. and Tinbergen, J. (1964). Linear polarization of the galactic background at 50 cm, *Bull. Astron. Inst. Neth.* **17**, 465–94.

Born, M. and Wolf, E. (1964). *Principles of Optics* (Pergamon Press, Oxford), 2nd edn (revised).

Borra, E.F., Landstreet, J.D. and Mestel, L. (1982). Magnetic stars, *Ann. Rev. Astron. Astrophys.* **20**, 191–220.

Bridle, A.H. and Perley, R.A. (1984). Extragalactic radio jets, *Ann. Rev. Astron. Astrophys.* **22**, 319–58.

Brouw, W.N. and Spoelstra, T.A.Th. (1976). Linear polarization of the galactic background at frequencies between 408 and 1411 MHz. Reductions, *Astron. Astrophys. Suppl.* **26**, 129–46.

Burn, B.J. (1966). On the depolarization of discrete radio sources by Faraday dispersion, *Mon. Not. Roy. Astron. Soc.* **133**, 67–83.

Casini, R. and Landi Degl'Innocenti, E. (1993). The polarized spectrum of hydrogen in the presence of electric and magnetic fields, *Astron. Astrophys.* **276**, 289–302.

Chandrasekhar, S. (1946). On the radiative equilibrium of a stellar atmosphere: XI, *Astrophys. J.* **104**, 110–32.

Chanmugam, G. (1992). Magnetic fields of degenerate stars, *Ann. Rev. Astron. Astrophys.* **30**, 143–84.

Chipman, R.A. (1989). Polarization analysis of optical systems, *Opt. Eng.* **28**, 90–9.

Chipman, R.A. (1992). The mechanics of polarization ray tracing, in *Polarization Analysis and Measurement*, eds. D.H. Goldstein and R.A. Chipman, Proc. SPIE **1746**, pp. 62–75 (SPIE, Bellingham).

Chipman, R.A. (1995). Mechanics of polarization ray tracing, *Opt. Eng.* **34**, 1636–45.

Chipman, R.A. and Chipman, L.J. (1989). Polarization aberration diagrams, *Opt. Eng.* **28**, 100–6.

Christiansen, W.N. and Högbom, J.A. (1985). *Radiotelescopes* (Cambridge University Press).

Clarke, D. (1974a). Nomenclature of polarized light, *Appl. Opt.* **13**, 3–5 and 222–4.

Clarke, D. (1974b) Polarimetric definitions, in *Planets, Stars and Nebulae Studied with Photopolarimetry*, ed. T. Gehrels (University of Arizona Press, Tucson).

Clarke, D. and Grainger, J.F. (1971). *Polarized Light and Optical Measurement* (Pergamon Press, Oxford).

Clarke, D. and Naghizadeh-Khouei, J. (1994). A reassessment of some polarization standard stars, *Astron. J.* **108**, 687–93.

Clarke, D. and Stewart, B.G. (1986). Statistical methods of stellar polarimetry, *Vistas Astron.* **29**, 27–51.

Clarke, D., Naghizadeh-Khouei, J., Simmons, J.F.L. and Stewart, B.G. (1993). A statistical assessment of zero-polarization catalogues, *Astron. Astrophys.* **269**, 617–26.

Clayton, G.C., Anderson, C.M., Magalhães, A.M., Code, A.D., Nordsieck, K.H., Meade, M.R., Wolff, M.J., Babler, B., Bjorkman, K.S., Schulte-Ladbeck, R.E., Taylor, M.J. and Whitney, B.A. (1992). The first spectropolarimetric study of the wavelength dependence of interstellar polarization in the ultraviolet, *Astrophys. J. Lett.* **385**, L53–7.

Clegg, R.E.S., Carter, D., Charles, P.A., Dick, J.S.B., Jenkins, C.R., King, D.L. and Laing, R.A. (1992). *ISIS Astronomers' Guide* (La Palma User Manual no. 22, Royal Greenwich Observatory, Cambridge).

Clemens, D.P., Leach, R.W., Barvainis, R. and Kane, B.D. (1990). Millipol, a mm/submm wavelength polarimeter: instrument, operation and calibration, *Pub. Astron. Soc. Pacific* **102**, 1064–76.

Coffeen, D.L. and Hansen, J.E. (1974), in *Planets, Stars and Nebulae Studied with Photopolarimetry*, ed. T. Gehrels (University of Arizona Press, Tucson).

Collett, E. (1993). *Polarized Light: Fundamentals and Applications,* (Marcel Dekker, New York).

Conway, R.G. (1974). Radio measurements of polarization, in *Planets, Stars and Nebulae Studied with Photopolarimetry*, ed. T. Gehrels (University of Arizona Press, Tucson).

Costa, E., Cinti, M.N., Feroci, M., Matt, G. and Rapisarda, M. (1994). Scattering polarimetry for X-ray astronomy by means of scintillating fibers, in *X-Ray and Ultraviolet Polarimetry*, ed. S. Fineschi, Proc. SPIE **2010**, pp. 45–56 (SPIE, Bellingham).

Crangle, J. and Gibbs, M. (1994). Units and unity in magnetism: a call for consistency, *Physics World*, Nov. 1994, pp. 31–2 (and readers' letters in subsequent issues).

Deguchi, S. and Watson, W.D. (1985). Circular polarization of interstellar absorption lines at radio frequencies, *Astrophys. J.* **289**, 621–9.

Donati, J.-F., Semel, M. and Rees, D.E. (1992). Circularly polarized spectroscopic observations of RS CVn systems, *Astron. Astrophys.* **265**, 669–81.

Duerbeck, H.W. and Schwarz, H.E. (1995). A high resolution spectrum of the symbiotic nova RR Telescopii, to be submitted to *Astron. Astrophys. Suppl.*

Dulk, G.A. (1985). Radio emission from the sun and stars, *Ann. Rev. Astron. Astrophys.* **23**, 169–224.

Elmore, D.F. (1990). A polarization calibration technique for the Advanced Stokes Polarimeter, National Center for Atmospheric Research Technical Note NCAR/TN-355+STR.

Elsner, R.F., Weisskopf, M.C., Novick, R., Kaaret, P. and Silver, E. (1990). On the performance of the scattering and crystal polarimeters for the SPECTRUM-X-Gamma mission, in *Polarimetry: Radar, Infrared, Visible, Ultraviolet and X-ray*, eds. R.A. Chipman and J.W. Morris, Proc. SPIE **1317**, pp. 372–81 (SPIE, Bellingham).

Engvold, O. (1992). LEST, the Large Earth-based Solar Telescope, *Europhys. News* **23**, 203–5.

Evans, D.L., Stofan, E.R., Jones, T.D. and Godwin, L.M. (1994). Earth from Sky, *Sci. Am.* Dec. 1994, pp. 44–9.

Favati, B., Landi Degl'Innocenti, E. and Landolfi, M. (1987). Resonance scattering of Lyman-α in the presence of an electric field, *Astron. Astrophys.* **179**, 329–38.

Fiebig, D. and Güsten, R. (1989). Strong magnetic fields in interstellar H_2O maser clumps, *Astron. Astrophys.* **214**, 333–8.

Fiebig, D., Wohlleben, R., Prata, A. and Rusch, W.V.T. (1991). Beam squint in axially symmetric reflector antennas with laterally displaced feeds, *IEEE Trans. Antennas and Propagation* **AP-39**, 774–9.

Fineschi, S., ed. (1994) *X-Ray and Ultraviolet Polarimetry*, Proc. SPIE **2010** (SPIE, Bellingham).

Flett, A.M. and Murray, A.G. (1991). First results from a submillimeter polarimeter on the James Clerk Maxwell Telescope, *Mon. Not. Roy. Astron. Soc.* **249**, 4p–6p.

Fomalont, E.B. and Perley, R.A. (1989). Calibration and Editing, in *Synthesis Imaging in Radio Astronomy*, eds. R.A. Perley, F.R. Schwab and A.H. Bridle (ASP Conference Series **6**, Astronomical Society of the Pacific, San Francisco).

Fosbury, R., Cimatti, A. and di Serego Alighieri, S. (1993). The limits of faint-object polarimetry, *ESO Messenger* no 74, pp. 11–15.

García-Barreto, J.A., Burke, B.F., Reid, M.J., Moran, J.M., Haschick, A.D. and Schilizzi, R.T. (1988). Magnetic field structure of the star-forming region W3(OH): VLBI spectral line results, *Astrophys. J.* **326**, 954–66.

Gehrels, T. (1974). Introduction and overview, in *Planets, Stars and Nebulae Studied with Photopolarimetry*, ed. T. Gehrels (University of Arizona Press, Tucson).

Gehrels, T. and Teska, T.M. (1960). A Wollaston photometer, *Proc. Astron. Soc. Pacific* **72**, 115–22.

Goodrich, R.W. (1991). High-efficiency 'superachromatic' polarimetry optics for use in optical astronomical spectrographs, *Pub. Astron. Soc. Pacific* **103**, 1314–22.

Hagyard, M.J., ed. (1985). *Measurements of Solar Vector Magnetic Fields*, NASA Conference Publication 2374 (NASA Scientific and Technical Information Branch)

Hamaker, J.P., Bregman, J.D. and Sault, R.J. (1995). Understanding radio polarimetry: I. Mathematical foundations, accepted by *Astron. Astrophys. Suppl.* for 1996.

Hecht, E. and Zajac, A. (1974). *Optics* (Addison-Wesley, Reading, MA).

Heiles, C. (1989). Magnetic fields, pressures and thermally unstable gas in prominent H I shells, *Astrophys. J.* **336**, 808–21.

Heiles, C. and Troland, T.H. (1982). Measurements of magnetic field strengths in the vicinity of Orion, *Astrophys. J. Lett.* **260**, L23–6.

Hildebrand, R.H. (1988). Magnetic fields and stardust, *Quart. J. Roy. Astron. Soc.* **29**, 327–51.

Hildebrand, R.H., Dragovan, M. and Novak, G. (1984). Detection of submillimeter polarization in the Orion Nebula, *Astrophys. J. Lett.* **284**, L51–4.

Hough, J.H., Peacock, T. and Bailey, J.A. (1991). A multiband two-beam optical and infrared polarimeter, *Mon. Not. Roy. Astron. Soc.* **248**, 74–8.

Hough, J.H., Chrysostomou, A. and Bailey, J.A. (1994). A new imaging infrared polarimeter, in *Infrared Astronomy with Arrays: The Next Generation*, ed. I.S. McLean, Astrophysics and Space Science Library **190** (Kluwer Academic Publishers, Dordrecht), pp. 287–9.

Hovenier, J.W. (1994). Structure of a general pure Mueller matrix, *Appl. Opt.* **33**, 8318–24.

Hovenier, J.W. and Van der Mee, C.V.M. (1983). Fundamental relationships relevant to the transfer of polarized light in a scattering atmosphere, *Astron. Astrophys.* **128**, 1–16.

IAU (1973). Polarization Definitions, *Trans. Intern. Astron. Union* **XVB**, 166.

IRE (1942). Standards on polarization concepts and terminology, *Proc. Inst. Radio Eng.* **30**, no 7, part III, suppl. IW47.

IEEE (1969). IEEE Standard Definitions of Terms for Radio Wave Propagation, *IEEE Trans. Antennas and Propagation* **AP-17**, 270–5.

IEEE (1972). Right-handed (clockwise) polarized wave (radio wave propagation), in *IEEE Standard Dictionary of Electrical and Electronic Terms*, (IEEE, New York) p. 500.

Jenkins, F.A. and White, H.E. (1950). *Fundamentals of Optics* (McGraw-Hill, New York).

Jerrard, H.G. (1954). Transmission of light through birefringent and optically active media: the Poincaré sphere, *J. Opt. Soc. Am.* **44**, 634–40.

Johnson, J.J. and Jones, T.J. (1991). From red giant to planetary nebula: dust, asymmetry and polarization, *Astron. J.* **101**, 1735–51.

Jones, R.C. (1941). A new calculus for the treatment of optical systems: I. Description and discussion of the calculus, *J. Opt. Soc. Am.* **31**, 488–93.

Jones, T.W. (1988). Polarization as a probe of magnetic field and plasma properties of compact radio sources: simulation of relativistic jets, *Astrophys. J.* **332**, 678–95.

Jones, T.W. and O'Dell, S.L. (1977). Transfer of polarized radiation in self-absorbed synchrotron sources. I. Results for a homogeneous source, *Astrophys. J.* **214**, 522–39.

Kaaret, P., Novick, R., Martin, C., Hamilton, T., Sunyaev, R., Lapshov, I., Silver, E., Weisskopf, M., Elsner, R., Chanan, G., Manzo, G., Costa, E., Fraser, G. and Perola, G.C. (1989). SXRP: a focal plane stellar X-ray polarimeter for the SPECTRUM-X-Gamma mission, in *X-ray/EUV Optics for Astronomy and Microscopy*, ed. R.B. Hoover, Proc. SPIE **1160**, pp. 587–97 (SPIE, Bellingham).

Kaaret, P., Schwartz, J., Soffitta, P., Dwyer, J., Shaw, P., Hanany, S., Novick, R., Sunyaev, R., Lapshov, I., Silver, E., Ziock, K.P., Weisskopf, M.C., Elsner, R.F., Ramsey, B., Costa, E., Rubini, A., Feroci, M., Piro, L., Manzo, G., Giarrusso, S., Santangelo, A., Scarsi, L., Perola, G.C., Massaro, E. and Matt, G. (1994). Status of the Stellar X-Ray Polarimeter for the Spectrum-X-Gamma Mission, in *X-Ray and Ultraviolet Polarimetry*, ed. S. Fineschi, Proc. SPIE **2010**, pp. 22–7 (SPIE, Bellingham).

Kattawar, G.W., Plass, G.N. and Hitzfelder, S.J. (1976). Multiple scattered radiation emerging from Rayleigh and continental haze layers. 1: Radiance, polarization and neutral points, *Appl. Opt.* **15**, 632–47.

Keller, C.U. and von der Lühe, O. (1992). Solar speckle polarimetry, *Astron. Astrophys.* **261**, 321–8.

Keller, C.U., Aebersold, F., Egger, U., Povel, H.P., Steiner, P. and Stenflo, J.O. (1992). Zürich imaging Stokes polarimeter ZIMPOL I – design review, *LEST Technical Report* no. 53 (Institute of Theoretical Astrophysics, University of Oslo).

Kemp, J.C. (1969). Piezo-optical birefringence modulators: new use for a long-known effect, *J. Opt. Soc. Am.* **59**, 950–4. Reprinted in *Selected Papers on Polarimetry*, ed. B.H. Billings, SPIE Milestone series **MS23** (SPIE, Bellingham, 1990).

Klaas, U., Krüger, H., Heinrichsen, I., Heske, A. and Laureys, R. (1994). *Isophot Observers' Manual* (ESA, Noordwijk).

Kliger, D.S, Lewis, J.W. and Randall, C.E. (1990). *Polarized Light in Optics and Spectroscopy* (Academic Press, San Diego).

Koester, C.J. (1959). Achromatic combinations of half-wave plates, *J. Opt. Soc. Am.* **49**, 405–9. Reprinted in *Selected Papers on Polarimetry*, ed. B.H. Billings, SPIE Milestone series **MS23** (SPIE, Bellingham, 1990).

Können, G.P. and Tinbergen, J. (1991). Polarimetry of a 22° halo, *Appl. Opt.* **30**, 3382–400.

Können, G.P., Schoenmaker, A.A. and Tinbergen, J. (1993). A polarimetric search for ice crystals in the upper atmosphere of Venus, *Icarus* **102**, 62–75.

Korhonen., T., Piirola, V. and Reiz, A. (1984). Polarization measurements at La Silla, *ESO Messenger* no. 38, pp. 20–4.

Kotov, Yu.D. (1988). Methods of measurement of gamma-ray polarization, *Space Sci. Rev.* **49**, 185–95.

Kraus, J.D. (1966) *Radio Astronomy* (McGraw-Hill, New York).

Kronberg, P.P. (1994). Extragalactic magnetic fields, *Rep. Prog. Phys.* **57**, 325–82.

Landau, L.D. and Lifshitz, E.M. (1975). *The Classical Theory of Fields* (Pergamon Press, Oxford), 4th edn (revised, in English).

Landi Degl'Innocenti, E. (1987). Transfer of polarized radiation, using 4×4 matrices, in *Numerical Radiative Transfer*, ed. W. Kalkofen (Cambridge University Press).

Landi Degl'Innocenti, E. (1992). Magnetic field measurements, in *Solar Observations: Techniques and Interpretation*, eds. F. Sánchez, M. Collados and M. Vázquez, First Canary Islands Winter School of Astrophysics (Cambridge University Press).

Landi Degl'Innocenti, E. and Landi Degl'Innocenti, M. (1981). Radiative transfer for polarized radiation: symmetry properties and geometrical interpretation, *Nuovo Cimento* **62B**, 1–16.

Landstreet, J.D. (1992). Magnetic fields at the surfaces of stars, *Astron. Astrophys. Rev.* **4**, 35–77.

Lang, K.R. (1980). *Astrophysical Formulae* (Springer, Berlin).

Léna, P. (1988). *Observational Astrophysics* (Springer, Berlin).

Leroy, J.L. (1985). The Hanle effect applied to magnetic field measurements, in *Measurements of Solar Vector Magnetic Fields*, ed. M.J. Hagyard (NASA Conference publication 2374, NASA Scientific and Technical Information Branch).

Liebert, J., Schmidt, G.D., Lesser, M., Stepanian, J.A., Lipovetsky, V.A., Chaffee, F.H., Foltz, C.B. and Bergeron, P. (1994). Discovery of a dwarf carbon star with a white dwarf companion and of a highly magnetic degenerate star, *Astrophys. J.* **421**, 733–7.

McClain, S.C., Hillman, L.W. and Chipman, R.A. (1993). Polarization ray tracing in anisotropic optically active media. *J. Opt. Soc. Am.* A **10**, 2371–93.

McGuire, J.P. and Chipman, R. (1988). Polarization accuracy of LEST, *LEST Technical Report* no. 36 (Institute of Theoretical Astrophysics, University of Oslo).

McKinnon, M.M. (1992a). Point source polarization calibration of a phased array, *Astron. Astrophys.* **260**, 533–42.

McKinnon, M.M. (1992b). Instrumental origin of a $180°$ phase shift in polarization data, *Astron. Astrophys.* **266**, 113–16.

Maronna, R., Feinstein, C. and Clocchiatti, A. (1992). Optimal estimation of Stokes' parameters, *Astron. Astrophys.* **260**, 525–32.

Martin, P.G. and Whittet, D.C.B. (1990). Interstellar extinction and polarization in the infrared, *Astrophys. J.* **357**, 113–24.

Martínez Pillet, V. and Sánchez Almeida, J. (1991). A proposal for a low instrumental polarization coudé telescope, *Astron. Astrophys.* **252**, 861–5.

Matt, G., Costa, E., Perola, G.C. and Piro, L. (1989). X-ray polarization of the reprocessed emission from accretion disk in Seyfert galaxies, in *Proc. 23rd ESLAB Symposium, Bologna* **ESA SP-296**, pp. 991–3.

Mészáros, P., Novick, R., Chanan, G.A., Weisskopf, M.C. and Szentgyörgyi, A. (1988). Astrophysical implications and observational prospects of X-ray polarimetry, *Astrophys. J.* **324**, 1056–67.

Michel, F.C. (1991) *Theory of Neutron Star Magnetospheres* (University of Chicago Press).

Minchin, N.R, Hough, J.H., McCall, A., Burton, M.G., McCaughrean, M.J., Aspin, C., Bailey, J.A., Axon, D.J. and Sato, S. (1991). Near-infrared imaging polarimetry of bipolar nebulae – I. The BN-KL region of OMC-1, *Mon. Not. Roy. Astron. Soc.* **248**, 715–29.

Mishchenko, M.I., Lacis, A.A. and Travis, L.D. (1994). Errors induced by the neglect of polarization in radiance calculations for Rayleigh-scattering atmospheres, *J. Quant. Spectro. Rad. Trans.* **51**, 491–510.

Morgan, M.F., Chipman, R.A. and Torr, D.G. (1990). An ultraviolet polarimeter for characterization of an imaging spectrometer, in *Polarimetry: Radar, Infrared, Visible, Ultraviolet and X-ray*, eds. R.A. Chipman and J.W. Morris, Proc. SPIE **1317**, pp. 384–94 (SPIE, Bellingham).

Morris, D., Radhakrishnan, V. and Seielstad, G.A. (1964). On the measurement of polarization distributions over radio sources, *Astrophys. J.* **139**, 551–9.

Morsi, H.W. and Reich, W. (1986). A new 32 GHz radio continuum receiving system for the Effelsberg 100-m telescope, *Astron. Astrophys.* **163**, 313–20.

Mueller, H. (1948). The foundations of optics (abstract of conference paper), *J. Opt. Soc. Am.* **38**, 661.

Murdin, P.G. (1990). Wood's anomalies in the INT spectrograph, La Palma Technical Note no. 76 (Royal Greenwich Observatory, Cambridge).

Murray, A.G., Flett, A.M., Murray, G. and Ade, P.A.R. (1992) High efficiency half-wave plates for submillimetre polarimetry, *Infrared Phys.* **33**, 113–25.

Myers, P.C. and Goodman, A.A. (1988). Evidence for magnetic and virial equilibrium in molecular clouds, *Astrophys. J. Lett.* **326**, L27–30.

Napier, P.J. (1989). The primary antenna elements, in *Synthesis Imaging in Radio Astronomy*, eds. R.A. Perley, F.R. Schwab and A.H. Bridle (ASP Conference Series **6**, Astronomical Society of the Pacific, San Francisco).

Neininger, N. (1992). The magnetic field structure of M51, *Astron. Astrophys.* **263**, 30–6.

Nordsieck, K.H, Code, A.D., Anderson, C.M., Meade, M.R., Babler, B., Michalski, D.E., Pfeifer, R.H. and Jones, T.E. (1994a). Exploring ultraviolet astronomical polarimetry: results from the Wisconsin Ultraviolet Photo-Polarimeter Experiment (WUPPE), in *X-Ray and Ultraviolet Polarimetry*, ed. S. Fineschi, Proc. SPIE **2010**, pp. 2–11 (SPIE, Bellingham).

Nordsieck, K.H, Marcum, P., Jaehnig, K.P. and Michalski, D.E. (1994b). New techniques in ultraviolet astronomical polarimetry: wide-field imaging and far ultraviolet spectropolarimetry, in *X-Ray and Ultraviolet Polarimetry*, ed. S. Fineschi, Proc. SPIE **2010**, pp. 28–36 (SPIE, Bellingham).

November, L.J (1989). Determination of the Jones matrix for the Sacramento Peak vacuum tower telescope, *Opt. Eng.* **28**, 107–13.

Pancharatnam, S. (1955a). Achromatic combinations of birefringent plates. Part I. An achromatic circular polarizer, *Proc. Indian Acad. Sci.* **A41**, 130–6.

Pancharatnam, S. (1955b). Achromatic combinations of birefringent plates. Part II. An achromatic quarter-wave plate, *Proc. Indian Acad. Sci.* **A41**, 137–44. Reprinted in *Selected Papers on Polarimetry*, ed. B.H. Billings, SPIE Milestone series **MS23** (SPIE, Bellingham, 1990).

Peters, C.J. (1964). Light depolarizer, *Appl. Opt.* **3**, 1502–3.

Piirola, V. (1973). A double image chopping polarimeter, *Astron. Astrophys.* **27**, 383–8.

Povel, H., Aebersold, H. and Stenflo, J.O. (1989) CCD image sensor as a demodulator in a 2-D polarimeter with a piezo-elastic modulator, *LEST Technical Report* no. 40 (Institute of Theoretical Astrophysics, University of Oslo).

Ramachandran, G.N. and Ramaseshan, S. (1952). Magneto-optic rotation in birefringent media – Application of the Poincaré sphere. *J. Opt. Soc. Am.* **42**, 49–56. Reprinted in *Selected Papers on Polarimetry*, ed. B.H. Billings, SPIE Milestone series **MS23** (SPIE, Bellingham, 1990).

Rand, R.J. and Kulkarni, S.R. (1989). The local Galactic magnetic field, *Astrophys. J.* **343**, 760–72.

Rees, D.E. (1987). A gentle introduction to polarized radiative transfer, in *Numerical Radiative Transfer*, ed. W. Kalkofen (Cambridge University Press).

Rutten, R.G.M. and Dhillon, V.S. (1992). Spectropolarimetry of the nova-like variable RW Trianguli, *Astron. Astrophys.* **253**, 139–44.

Saikia, D.J. and Salter, C.J. (1988). Polarization properties of extragalactic radio sources, *Ann. Rev. Astron. Astrophys.* **26**, 93–144.

Sánchez Almeida, J. (1992). Radiative transfer for polarized light: equivalence between Stokes parameters and coherency matrix formalisms, *Solar Phys.* **137**, 1–14.

Sánchez Almeida, J. (1994). Instrumental polarization in the focal plane of telescopes. II. Effects induced by seeing, *Astron. Astrophys.* **292**, 713–21.

Sánchez Almeida, J. and Martínez Pillet, V. (1992). Instrumental polarization in the focal plane of telescopes, *Astron. Astrophys.* **260**, 543–55.

Sault, R.J., Hamaker, J.P. and Bregman, J.D. (1995). Understanding radio polarimetry: II. Instrumental calibration of an interferometer array, accepted by *Astron. Astrophys. Suppl.* for 1996.

Scarrott, S.M. (1991). Optical polarization studies of astronomical objects, *Vistas Astron.* **34**, 163–77.

Scarrott, S.M., Warren-Smith, R.F., Pallister, W.S., Axon, D.J. and Bingham, R.G. (1983). Electronographic polarimetry: the Durham polarimeter, *Mon. Not. Roy. Astron. Soc.* **204**, 1163–77.

Schild, H. and Schmid, H.M. (1992). Raman scattering, spectropolarimetry and symbiotic stars, *Gemini* (Royal Greenwich Observatory newsletter) no. 37, pp. 4–7.

Schmidt, A., Wongsowijoto, S., Lochner, O., Reich, W., Reich, P., Fürst, E. and Wielebinski, R. (1993). *Max-Planck-Inst. f. Radioastronomie, Bonn* Technischer Bericht no. 73.

Schmidt, G.D., Elston, R. and Lupie, O.L. (1992a). The Hubble Space Telescope northern-hemisphere grid of stellar polarimetric standards, *Astron. J.* **104**, 1563–7.

Schmidt, G.D., Stockman, H.S. and Smith, P.S. (1992b). Discovery of a sub-megagauss magnetic white dwarf through spectropolarimetry, *Astrophys. J. Lett.* **398**, L57–L60.

Schüssler, M. and Schmidt, W., eds. (1994). *Solar Magnetic Fields* (Cambridge University Press).

Schwarz, U.J., Troland, T.H., Albinson, J.S., Bregman, J.D., Goss, W.M. and Heiles, C. (1986). Aperture synthesis observations of the 21 centimeter Zeeman effect towards Cassiopeia A, *Astrophys. J.* **301**, 320–30.

Scott, P.F. and Ryle, M. (1977). A rapid method for measuring the figure of a radio telescope reflector, *Mon. Not. Roy. Astron. Soc.* **178**, 539–45.

Semel, M. (1987). Polarimetry and imagery through uniaxial crystals. Application to solar observations with high spatial resolution, *Astron. Astrophys.* **178**, 257–62.

Serkowski, K., Mathewson, D.S. and Ford, V.L. (1975). Wavelength dependence of interstellar polarization and ratio of total to selective extinction, *Astrophys. J.* **196**, 261–90.

Shurcliff, W.A. (1962). *Polarized Light, Production and Use* (Harvard University Press, Cambridge, MA).

Silver, E., Simionivici, A., Labov, S., Novick, R., Kaaret, P., Martin, C., Hamilton, T., Weisskopf, M., Elsner, R., Beeman, J., Chanan, G., Manzo, G., Costa, E., Perola, G. and Fraser, G. (1989). Bragg crystal polarimeters, in *X-ray/EUV Optics for Astronomy and Microscopy*, ed. R.B. Hoover, Proc. SPIE **1160**, pp. 598–609 (SPIE, Bellingham).

Simmons, J.W. and Guttmann, M.J. (1970). *States, Waves and Photons: A Modern Introduction to Light* (Addison-Wesley, Reading, MA).

Simon, R. (1990). Nondepolarizing systems and degree of polarization, *Opt. Commun.* **77**, 349–54.

Smith, C.H., Aitken, D.K. and Moore, T.J.T. (1994). An imaging polarimeter for the mid-infrared, in *Instrumentation in Astronomy VIII*, eds. D.R. Crawford and E.R. Craine, Proc SPIE **2198**, 736–43 (SPIE, Bellingham).

Smith, F.G., Jones, D.H.P., Dick, J.S.B. and Pike, C.D. (1988). The optical polarization of the Crab pulsar, *Mon. Not. Roy. Astron. Soc.* **233**, 305–19.

Snell, J.F. (1978). Radiometry and photometry, in *Handbook of Optics*, eds. W.G. Driscoll and W. Vaughan (McGraw-Hill, New York).

Sofue, Y., Fujimoto, M. and Wielebinski, R. (1986). Global structure of magnetic fields in spiral galaxies, *Ann. Rev. Astron. Astrophys.* **24**, 459–97.

Somerville, W.B., Allen, R.G., Carnochan, D.J., He, L., McNally, D., Martin, P.G., Morgan, D.H., Nandy, K., Walsh, J.R., Whittet, D.C.B., Wilson, R. and Wolff, M.J. (1994). Ultraviolet interstellar polarization observed with the Hubble Space Telescope, *Astrophys. J. Lett.* **427**, L47–50.

Spoelstra, T.A.Th. (1972a). A survey of linear polarization at 1415 MHz. I. Method of reduction and results for the North Polar Spur, *Astron. Astrophys. Suppl.* **5**, 205–38.

Spoelstra, T.A.Th. (1972b). A survey of linear polarization at 1415 MHz. III. Method of reduction and results for the galactic spurs, *Astron. Astrophys. Suppl.* **7**, 169–230.

Spoelstra, T.A.Th. (1984). Linear polarization of the galactic radio emission at frequencies between 408 and 1411 MHz. II. Discussion, *Astron. Astrophys.* **135**, 238–48.

Spoelstra, T.A.Th. (1992). Mapping of continuum polarization, *Neth. Found. Res. Astron.*, note 604.

Stammes, P., de Haan, J.F. and Hovenier, J.W. (1989). The polarized internal radiation field of a planetary atmosphere, *Astron. Astrophys.* **225**, 239–59.

Stenflo, J.O. (1984). Solar magnetic and velocity-field measurements: new instrument concepts, *Appl. Opt.* **23**, 1267–78.

Stenflo, J.O. (1989). Small-scale magnetic structures on the Sun, *Astron. Astrophys. Rev.* **1**, 3–48.

Stenflo, J.O. (1994). *Solar Magnetic Fields* (Kluwer Academic Publishers, Dordrecht).

Stenflo, J.O., Keller, C.U. and Povel, H. (1992). Demodulation of all four Stokes parameters with a single CCD: ZIMPOL II – conceptual design, *LEST Technical Report* no. 54 (Institute of Theoretical Astrophysics, University of Oslo).

Stinebring, D.R., Cordes, J.M., Rankin, J.M., Weisberg, J.M. and Boriakoff, V. (1984). Pulsar polarization fluctuations, I. 1404 MHz statistical summaries, *Astrophys. J. Suppl.* **55**, 247–77, II. 800 MHz statistical summaries, *Astrophys. J. Suppl.* **55**, 279–88.

Stokes, G.G. (1852). On the composition and resolution of streams of polarized light from different sources, *Trans. Camb. Phil. Soc.* **9**, part III, pp. 399–416. Reprinted in *Mathematical and Physical Papers* **3**, 233–58 (Cambridge University Press, 1901).

Taylor, J.H. and Stinebring, D.R. (1986). Recent progress in the understanding of pulsars, *Ann. Rev. Astron. Astrophys.* **24**, 285–327.

Thompson, A.R., Moran, J.M. and Swenson, G.W. (1986). *Interferometry and Synthesis in Radio Astronomy* (John Wiley & Sons, New York).

Tinbergen, J. (1972). Achromatic polarization modulators for multichannel polarimeters, in *ESO/CERN Conference on Auxiliary Instrumentation for Large Telescopes*, eds. S. Laustsen and A. Reiz (European Southern Observatory, Garching bei München).

Tinbergen, J. (1973). Precision spectropolarimetry of starlight: development of a wide-band version of the Dollfus polarization modulator, *Astron. Astrophys.* **23**, 25–48.

Tinbergen, J. (1974). Application of Mueller calculus in astronomical polarimetry: achromatic modulators and polarisation converters, and depolarizers, in *Planets, Stars and Nebulae Studied with Photopolarimetry*, ed. T. Gehrels (University of Arizona Press, Tucson).

Tinbergen, J. (1987a). Polarisation fidelity of telescopes: necessary condition for high accuracy, in *Observational Astrophysics with High-Precision Data*, 27th Liège Symposium (Institut d'Astrophysique, Université de Liège).

Tinbergen, J. (1987b). *User Guide to the Multi-Purpose Fotometer (MPF)* La Palma User Manual no. 14 (Royal Greenwich Observatory, Cambridge).

Tinbergen, J. (1988). Observational errors caused by polarisation effects in folded telescopes and instruments, in *ESO Conference on Very Large Telescopes and their Instrumentation*, ed. M.-H. Ulrich (European Southern Observatory, Garching bei München).

Tinbergen, J. (1995). Array polarimetry and optical-differencing photometry, in *New Developments in Array Technology and Applications*, ed. A.G. Davis Philip, IAU Symposium no 167 (Kluwer Academic Publishers, Dordrecht).

Tinbergen, J. and Rutten, R.G.M. (1992). *A User Guide to WHT Spectropolarimetry*, La Palma User Manual no. 21 (Royal Greenwich Observatory).

Trammell, S.R., Dinerstein, H.L. and Goodrich, R.W. (1994). Evidence for the early onset of aspherical structure in the planetary nebula formation process: spectropolarimetry of post-AGB stars, *Astron. J.* **108**, 984–97.

Troland, T.H. and Heiles, C. (1982a). The Zeeman effect in 21 centimeter line radiation: methods and initial results, *Astrophys. J.* **252** , 179–92.

Troland, T.H. and Heiles, C. (1982b). Magnetic field measurements in two expanding H I shells. *Astrophys. J. Lett.* **260** L19–22.

Tsunemi, H., Hayashida, K., Tamura, K., Nomoto, S., Wada, M., Miyata, E. and Miura, N. (1994). Application of a charge-coupled device as an X-ray polarimeter, in *X-Ray and Ultraviolet Polarimetry*, ed. S. Fineschi, Proc. SPIE **2010**, pp. 201–10 (SPIE, Bellingham).

Turlo, Z., Forkert, T., Sieber, W. and Wilson, W. (1985). Calibration of the instrumental polarization of radio telescopes, *Astron. Astrophys.* **142**, 181–8.

Van de Hulst, H.C. (1957). *Light Scattering by Small Particles* (Wiley, New York). Reprinted 1981 (Dover, New York).

Van de Hulst, H.C. (1980). *Multiple Light Scattering* (Academic Press, New York).

Verschuur, G.L. (1989). Measurements of the 21 centimeter Zeeman effect in high-latitude directions, *Astrophys. J.* **339**, 163–70.

Verschuur, G.L. (1995a). Zeeman effect observations of H-I emission features: 1. Magnetic field limits for three regions based on observations corrected for polarized beam structure, *Astrophys. J.*, **451**, in press.

Verschuur, G.L. (1995b). Zeeman effect observations of H-I emission features: 2. Results of an attempt to confirm previous claims of field detections, *Astrophys. J.*, **451**, in press.

Volkmer, R. (1994). Two-dimensional high resolution solar spectro-polarimetry, in *Solar Magnetic Fields*, eds. M. Schüssler and W. Schmidt (Cambridge University Press).

Voshchinnikov, N.V. and Karjukin, V.V. (1994). Multiple scattering of polarized radiation in circumstellar dust shells, *Astron. Astrophys.* **288**, 883–96.

Walther, D.M., Aspin, C.A. and McLean, I.S. (1990). The exciting star in G35.2N, *Astrophys. J.* **356**, 544–8.

Weiler, K.W. (1973). The synthesis radio telescope at Westerbork: methods of polarization measurement, *Astron. Astrophys.* **26**, 403–7.

Weisskopf, M.C., Silver, E.H., Kestenbaum, H.L., Long, K.S. and Novick, R. (1978). A precision measurement of the X-ray polarization of the Crab Nebula without pulsar contamination, *Astrophys. J. Lett.* **220**, L117–21.

Westerhout, G., Seeger, C.L., Brouw W.N. and Tinbergen, J. (1962). Polarization of the galactic 75-cm radiation, *Bull. Astron. Inst. Neth.* **16**, 187–212.

Whittet, D.C.B., Martin, P.G., Hough, J.H., Rouse, M.F., Bailey, J.A. and Axon, D.J. (1992). Systematic variations in the wavelength dependence of interstellar linear polarization, *Astrophys. J.* **386**, 562–77.

Wielebinski, R. and Krause, F. (1993). Magnetic fields in galaxies, *Astron. Astrophys. Rev.* **4**, 449–85.

Wieringa, M.H., de Bruyn, A.G., Jansen, D., Brouw, W.N. and Katgert, P. (1993). Small scale polarization structure in the diffuse galactic emission at 325 MHz, *Astron. Astrophys.* **268**, 215–29.

Xilouris, K.M. (1991). Pulsars as instrumental polarization calibrators, *Astron. Astrophys.* **248**, 323–7.

Index